The Periodic Table
EXPERIMENT & THEORY

Other titles by
J. S. F. Pode
written in collaboration
with G. F. Liptrot

A Basic Course in Chemistry
Exploring Chemistry

The Periodic Table
EXPERIMENT & THEORY

J.S.F. Pode M.A. B.Sc.

A HALSTED PRESS BOOK

JOHN WILEY & SONS
New York

106941

First published 1971
by Mills & Boon Ltd
17–19 Foley Street, London W1A 1DR
Reprinted 1973

© J.S.F. Pode 1970

ISBN 0 470–69144–1
Library of Congress Number: 73–5338

Published in the U.S.A. by
Halsted Press, a Division of
John Wiley & Sons, Inc.
New York

Made and Printed in Great Britain
by Butler & Tanner Ltd, Frome and London

Contents

A•

Foreword

In this short text the author has accomplished exactly what he set out to do. He has first shown how the brilliant ideas of Mendeleev lead to a framework—the Periodic Table—which enables us to correlate the diverse facts of Chemistry; then he has employed these facts again, this time to show how they, and the framework itself, can be accounted for in terms of the modern ideas of structure and bonding.

Such a text is both an exciting new venture and a considerable achievement. I am sure the book will prove to be of great interest and help to both teachers and students of Chemistry.

R. O. C. Norman

University of York
June 1971

Preface

An introduction to the Periodic Table is an essential part of any course in chemistry. No student can survive for long on a diet of unconnected facts, impossible to organise and difficult to remember. The recognition of a pattern among individual pieces of data is a deeply satisfying achievement in itself, let alone the value of the pattern as a guide to the relative importance of the facts and as an aid to their memorisation.

Before the full impact of Mendeleev's discovery can be appreciated, however, the student must have an acquaintance with the immense diversity of experimental results available. A selection of information in a deliberately concise form, stressing regularities rather than anomalies, has been assembled in Part A of this book, and about 150 simple experiments have been included so that a wide variety of these unsupported statements can be verified. This material has been kept separate from its interpretation in terms of modern valency theory, which is the subject of Part B. Operational definitions are used for terms like 'electropositivity' when they occur in Part A, and the only concepts which are freely used are those of atomic weight and the nature of ions; molecular formulae are quoted without reference to the way in which the formulae have been confirmed. Apart from these exceptions Part A is an empirical account; words like 'covalent', 'valency' and 'oxidation state' are explicitly avoided, though the latter is implicit in the quotation of molecular formulae.

Each section of Part A is balanced by a similarly numbered section in Part B. The Introduction to Part A traces the coalescence of ideas about the nature of chemical families, while the Introduction to Part B deals with the emergence of the concept of orbitals a century later.

A.1 deals with the nature of a scientific law, the predictive successes of Mendeleev's system, its anomalies and failures, while B.1 shows how modern theoretical work has accounted for these but raised some new problems. A.2 deals with the operational concepts used in Part A, whereas B.2 reinterprets these in conceptual terms and elaborates further concepts for use in B.5. The best empirical shape of the modern Periodic Table is discussed in A.3 and its justification is given in B.3. A.4 and B.4 deal with Periodic trends of elements, hydrides, oxides and halides across the Periods.

A.5 and B.5 are a detailed consideration of the Groups. Separate headings are deliberately excluded from A.5 because they would differ for different groups and the section is primarily designed to show the sweep and unity of the Periodic System. A.5 has been kept very concise, even though it contains a wealth of detail, so that it may be read straight through at one sitting. More than anything else, such a reading will demonstrate to the student how the arrangement of such detail by Group and Period is essential for clarity before the more refined theoretical reasoning, as detailed in B.5, can be successfully applied to it.

A.6 and B.6 deal with the experimental evidence and theoretical justification for certain special topics, the diagonal relationship of the early Main Group elements for example.

The experiments at the end of Part A which support the factual material it contains have been chosen for their easy manipulation and unambiguous results. Unfortunately there is no possibility of anything approaching comprehensive coverage for reasons of space, and some elements like fluorine, beryllium and arsenic are far too dangerous for unsupervised work in the open laboratory despite their great chemical interest.

Part A of the book can stand on its own (though a diet of fact undiluted with theory tends to be somewhat arid) but it is no use pretending that it would take exactly its present form were it not for the theoretical promptings of Part B. This is particularly true of the empirical arrangement of the Periodic Table which occurs in A.3. On the other hand Part B of the book should certainly not be read in isolation from Part A. It is the author's belief that in recent years far too many little handbooks to modern valency theory have appeared which convey the impression that orbital theory is some kind of received truth about the universe rather than a set of elegant but essentially provisional hypotheses for making sense of the experimental results.

The way in which provisional hypotheses can become articles of faith is well shown by the 'octet rule' with its insistence on the immutable stability of the electronic configurations of the noble gases. For a long time the slavish insistence on this rule made the understanding of the bonding in electron deficient compounds like the boron hydrides artificially difficult, and the discovery of the compounds of xenon and krypton was delayed on doctrinal grounds. A similar danger is present in the use of the term 'electronic subshell' for the group of electrons which lie between one noble gas and the next. This term is misleading for two reasons: the word 'shell' suggests a static arrangement rather than a dynamic pattern, and secondly it summons up an image of a thin surface electron layer rather than a probability distribution which may be deeply penetrating. For this reason the term 'electronic subshell' has been replaced in this book by the term *rhythmic electron pattern*. The latter must of course be sharply distinguished from the total number of electrons associated with a particular principal quantum number, which is not of course closely related to the number of electrons between one noble gas and the next.

As Professor Medawar has often said, 'Science advances by a constant interplay between theory and experiment: that is what makes it a going concern.' This book is a deliberate attempt to give life to this interplay, to present science as it actually is without harping upon the specialist terms of scientific philosophy. The formulation of a scientific hypothesis is an imaginative act performed by one exceptional human being, and any writer who fails to do justice to this point of view is falsifying the nature of his subject. My hope is that no reader can fail to see after reading the book how greatly the insight of one man, Dmitri Mendeleev, has influenced chemical thinking, for it was he who gave meaning to the terms 'Group' and 'Period' which have been the cornerstones of all subsequent work.

J.S.F. Pode

Acknowledgements

First I should like to thank Dr. D. J. Waddington for much helpful advice about the structure of the book. Over the years Mr. D. Hughes has made me free of his great experience in designing the experiments contained in A.7, and I am grateful to Dr. G. F. Liptrot and Mr. J. Cook for checking some of these for me. Professor R. O. C. Norman was kind enough to spend some of his valuable time in writing the foreword, and Mrs. Kathleen Liptrot, as usual, translated a scruffy manuscript into elegant typescript. My greatest debt however is to my Scholarship Divisions at Eton, whose unfailing criticism of my wilder ideas about the presentation of theoretical chemistry was an essential refining process without which this book could never have emerged.

Introduction to Part A

Chemical families: historical background

The belief that the atoms of different elements had nothing in common with each other was one of the pillars of Dalton's atomic theory (1808). An attempt to suggest that all elements were formed by the coalescence of hydrogen atoms, on the flimsy and inaccurate evidence that all atomic weights were whole numbers, was made by the English doctor, Prout (1815). It was left to Berzelius to show that the atomic weight of chlorine was not 35, nor 36, but 35·5—as far from a whole number as it could possibly be. So perished the first attempt to find unity among the elements.

Yet it was undeniable that different atoms had very similar properties in some cases. The five following groups show obvious similarities.

Lithium, sodium, potassium. Melt below 200°C. Burn readily in air to oxides which cannot easily be reduced chemically. React violently with water to form strongly alkaline solutions. Form white salts which are stable to 600°C and almost invariably soluble in water.

Calcium, strontium, barium. Melt above 750°C. Burn freely in air to stable oxides. React with water to form strongly alkaline solutions. Form white salts stable to 600°C; the hydroxides are sparingly soluble in water, the sulphates and carbonates almost insoluble.

Sulphur, selenium, tellurium. Non-metals melting below 500°C. Form insoluble compounds by direct union with metals at red heat, which liberate vile smelling hydrides with acids. Form dioxides and trioxides which dissolve in water to give acidic solutions.

Chlorine, bromine, iodine. Non-metals boiling below 200°C, not very soluble in water. Combine vigorously with metals to form salts. The gaseous hydrides are displaced from their salts by high boiling point acids and these dissolve in water to give strongly acid solutions. Oxides are not readily formed.

Iron, cobalt, nickel. Metals with high boiling points which do not react with water. Form several insoluble oxides which react with acids to form coloured salts.

Men with a mathematical turn of mind will always play idly with numbers—even those on a church hymnboard during a dull sermon. In 1829 Dobereiner was the first to comment that the second atom in the 'triads' quoted above had an atomic weight which was always almost exactly the arithmetic mean of the atomic weights of the first and third elements. For example, $Li = 7$, $K = 39$ and $Na = (7 + 39)/2$ or 23; similarly $Cl = 35\cdot5$, $I = 127$ and $Br = (35\cdot5 + 127)/2$ or 80. But the differences between different members of the five triads were not uniform—indeed there was hardly any difference at all between iron, cobalt and nickel—and did not therefore appear to be significant. No further constructive work was possible until Cannizzaro's unambiguous atomic weight table in 1858.

Newlands (1863) was the first to see that if the known elements were written in ascending order of atomic weight, similar properties recurred in every eighth element, like notes in a musical scale. This system worked quite well for the lighter elements; for example it brought the Li/Na/K triad together. By failing to allow for enlarging the octave when dealing with the heavier elements the theory broke down and was greeted with derisive laughter at a meeting of the Chemical Society. The hostility behind the mirth arose not from the obvious imperfections in detail, rather because the idea that atomic properties recurred at fixed interval suggested that there was some sort of repeating structure inside the atom—a flat contradiction of the immensely successful atomic theory of Dalton.

From 1864 onwards the German chemist Lothar Meyer was accumulating evidence that certain quantitative properties of the elements were indeed periodic. The ratio of atomic weight/density of any solid element gives the relative volume of the atoms of that element. When these **'atomic volumes'** are plotted against atomic weight a curve is produced which reaches well-defined maxima at the atomic weights corresponding to the alkali metals (the modern form of this diagram is shown in Figure 1). However, not all atomic properties were periodic; the amount of heat required to raise one gramme atomic weight of any element—which contained the same number of atoms in each case—through $1°$ was not periodic, but constant, a truth entombed in the often erratic law of Dulong and Petit.

The time was ripe for a man of genius to gather together all these hints into a comprehensive statement. In 1869 Dmitri Mendeleev,

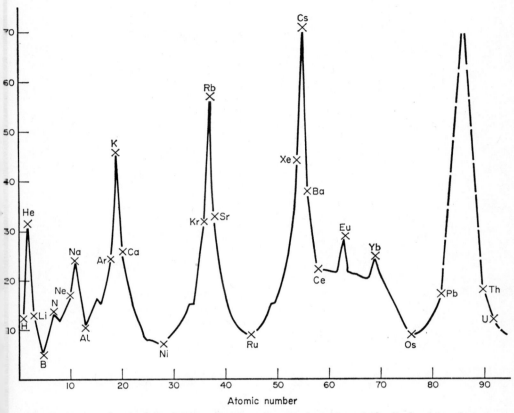

Fig. 1. Atomic volumes of the elements plotted against atomic numbers. (After Lothar Meyer.)

Professor of Chemistry at St. Petersburg, produced a detailed classification which included all the known elements, built round the twin concepts of '**Groups**' and '**Periods**'. The later discoveries of six noble gases, fourteen inner transitional elements and twelve man-made elements have led to modifications in the detail of Mendeleev's arrangement, but the use of Groups and Periods is just as important now as it was 100 years ago.

Part A: Experimental

A.1
The Periodic Law

A.1(1) STATEMENT OF THE PERIODIC LAW

The basic hypothesis underlying Mendeleev's classification may be
stated as follows:
"The chemical elements, as characterised by their physical and
chemical properties, fall into a repeating pattern if they are arranged
in order of increasing atomic weight."

A.1(2) THE FUNCTION OF A SCIENTIFIC HYPOTHESIS

The formulation of any hypothesis is an attempt by man to find
order in nature. There is no lumber room of the universe
containing a tablet of stone upon which is carved a Periodic Table;
Mendeleev's appreciation of a pattern among the atoms was an
imaginative act, an explanatory conjecture giving one of many
possible interpretations of a given set of facts. To formulate any
hypothesis involves a selection—in this case the abstraction of those
facts which are of fundamental significance from those which are
secondary. The choice of what is significant cannot be explained by
any logical means but is an imaginative act of scientific judgement.
For instance, in Mendeleev's initial scheme, it was important to
stress the similarities between the metal manganese and the gas
chlorine. Mendeleev had the insight to emphasise that both elements
form ions with the general formula XO_4^- while playing down the
gross dissimilarity of colour and thermal stability between ClO_4^-
and MnO_4^- as being of only secondary importance.

There are two further startling examples of Mendeleev's use of
scientific judgement to emphasise the periodic pattern. Firstly he
made the arbitrary decision that iron, cobalt and nickel were three
forms of the same metal so that they might more conveniently act
as a 'transition' between the A & B octaves in his enlarged Periods.
Secondly he flatly asserted that the atomic weights of titanium,
platinum and tellurium had been wrongly measured simply because
they did not fit happily into his periodic pattern—using the accepted
atomic weight values of the time. For instance, tellurium fell into the
halogen group whereas iodine its neighbour did not. Neither of

these decisions could be justified on any other grounds except that, in Mendeleev's judgement, they 'made sense'.

Any hypothesis is a kind of inspired guess, an attempt to visualise what the universe is like. If it leads to deductions which are later proved wrong by experiments designed to test them, it is false and must be discarded. On the other hand no hypothesis can be proved to be true; it is a provisional statement only and its ultimate validity is still in doubt even though it is fruitful and generates many predictions which are verifiable experimentally.

The provisional nature of hypotheses and the theories which are built on them is a constant spur to criticism and further experiment. The open questioning of accepted views is the very essence of science, which cannot therefore flourish under conditions of oppression and secrecy. Mendeleev himself was forced to resign his university post in Tsarist Russia.

A.1(3) PREDICTIVE SUCCESSES

Mendeleev was immediately able to correct the atomic weight of indium. The equivalent weight of indium was stated at that time to be 37·7; inaccurate experimental work on its specific heat had led to a value of $2 \times 37·7$ or 75·4 for its atomic weight. In Mendeleev's pattern there was no possible place for indium at this value, because between arsenic at 75 (probably M5) and bromine at 80 (certainly M7) there was already the element selenium, closely related to sulphur; clearly a divalent metal at 75·4 was hopelessly out of place. However there was a possible gap in the table for a trivalent metal between cadmium at 112 and tin at 118, and Mendeleev had no inhibitions about allotting an atomic weight of $3 \times 37·7$ or 113·1 to indium (the modern value for the atomic weight is 114·82). A similar readjustment was made for cerium a little later.

There were two gaps in Mendeleev's table between zinc and arsenic, and he had no hesitation in forecasting properties for such elements. For example, by adding together the atomic weights of elements above and below and on both sides of the unknowns, and dividing the result by 4, he arrived at probable values for the atomic weights.

PROBLEM: Using x and y for the atomic weights of the unknown elements duplicate Mendeleev's prediction of the possible atomic weights of these elements by constructing two simultaneous equations. Using books of reference make similar forecasts of the specific heats and the melting points of the chlorides.

Other physical and chemical properties can be approached in the same way. His hypothesis was triumphantly justified by the discovery of gallium (atomic weight 70) in 1875 and germanium (atomic weight 72) in 1886, with properties very close to those he had predicted.

The reclassification of beryllium, which was not finally accepted until 1890, was a further triumph. Until then, owing to an extraordinary similarity to aluminium (p. 51) and an anomalously low value for its specific heat which gave an erroneous calculation using Dulong and Petit's Law, the atomic weight of this element had been taken to be three times its equivalent weight, or 13·5. Mendeleev had firmly asserted that there was no room for it between carbon and nitrogen, and was justified by a measurement of a vapour density of 40 for beryllium chloride, whose formula thus proved to be $BeCl_2$ and led to an atomic weight of 9 for beryllium.

A.1(4) APPARENT FAILURES

(a) The comparative dissimilarities between A Groups and B Groups, which were classified together in the original scheme. For instance chromium (a metal of high m.p.) and selenium (a non-metal of low m.p.) were placed together, and so were manganese and chlorine. Despite similarities when both pairs of elements are showing their maximum combining power (chromates and selenates; permanganates and perchlorates) the differences between them were overwhelmingly more striking.
(b) As the determination of equivalent (and therefore atomic) weights became more accurate, certain definite anomalies besides the exchange of tellurium and iodine came to light. Potassium, atomic weight 39·1, is certainly not a noble gas, and argon, atomic weight 39·9, cannot masquerade as an alkali metal. Similar exchanges of Group occur with the pairs cobalt/nickel and thorium/protoactinium.
(c) The position of Mendeleev's three transitional metals as a kind of interlude between the A Groups and B Groups was clearly arbitrary and unsatisfactory.
(d) Mendeleev realised that there was something grievously wrong with his classification of the 'rare earths'. As late as 1891 he was postulating seventeen undiscovered elements, corresponding to a whole transitional series, between cerium, atomic weight 140, and ytterbium, atomic weight 173.

Nevertheless, despite these obvious imperfections, Mendeleev adhered fanatically to his hypothesis, maintaining that the apparent errors were of only secondary importance and would finally be explained. Events have of course proved him right: no gaps now remain to be filled, even though elements 43 and 61 do not occur in nature as they are radioactive and have very short half lives.

A.2
Empirical definitions

A.2(1) A GROUPS AND B GROUPS

There is no empirical distinction between elements in A Groups and elements in B Groups. Mendeleev, however, was forced to distinguish them on account of his hypothesis of the 'double octave' form of the Long Periods.

Similarly there is no clear cut empirical definition which serves to distinguish Main Group elements from transitional elements in the modern arrangement of the Periodic Table.

A.2(2) TRANSITIONAL ELEMENTS

The empirical properties of the transitional elements are detailed in A.6(9) but there is no short practicable definition.

The term **'transitional element'** was first used by Mendeleev in a quite different sense, to describe the three elements Fe, Co and Ni, which were intermediate between the A Group and B Group elements in the First Long Period. There was no empirical justification for this, indeed it was a clear case of giving a name to some phenomenon not fully understood in order to render it less puzzling: cf. "Thou shalt call his name Jehovah" in the Old Testament.

A.2(3) INNER TRANSITIONAL ELEMENTS

The most striking empirical properties of the inner transitional elements are their electropositive metallic nature, and the series of salts they form in which the equivalent weight of the metal is one third of its atomic weight.

Most of these elements were still undiscovered at the death of Mendeleev, partly because being so very similar they were difficult to separate, and partly because being strongly electropositive the pure metals are difficult to prepare from their oxides.

A.2(4) ELECTROPOSITIVITY

The more electropositive an element:
(a) The more violently it will react with water—provided there is no protective oxide layer on the surface to prevent this, as with aluminium.
(b) The more readily it will react with acid.
(c) The more thermally stable will be its salts.
(d) The less hydrolysed will be its salts in solution, and the less hydrated will be the crystals of these salts.
(e) The more basic will be its oxide and hydroxide.
(f) The fewer complexes will it form.

These properties are shown very well by potassium and the contrast between this very electropositive metal and the comparatively unreactive metal copper, which were classified as A and B elements in the same Group by Mendeleev, is vivid. Thus copper does not react with acids and water while potassium does so explosively. The oxide of potassium dissolves exothermically to give an alkaline solution whereas copper oxide is insoluble. Copper sulphate gives off water of crystallisation on gentle heating and turns black if roasted strongly, and its solution gives a strongly acid reaction to indicators; potassium sulphate does none of these things. Copper sulphate solution first forms a precipitate with ammonium hydroxide but this soon dissolves with the formation of the complex ion $Cu(NH_3)_4{}^{2+}$; potassium sulphate solution is unaffected.

A.2(5) VALENCY

'The number of atoms of hydrogen with which one atom of the element will combine.' This definition justifies the assumption that the valency of an element is a small whole number, since it is obviously not possible to have a fraction of a hydrogen atom. From an empirical point of view it is very unsatisfactory in that a large number of elements do not form well established compounds with hydrogen.

As discussed in B.3(5) the word valency is ambiguous and is best discarded.

A.2(6) SALT-LIKE COMPOUNDS

These are made by the action of strong acids on the oxides or
hydroxides of electropositive metals. They have two outstanding
characteristics:
(a) They melt before they decompose, and the molten salt conducts
electricity.
(b) They dissolve in water, and their solutions conduct electricity.

It was known that there were certain exceptions to these rules:
barium sulphate is insoluble and infusible for example.

Circumstances which favoured salt-like properties were investigated
by Fajans. He found that large cations of low charge combined with
small anions were the most favourable (his assessment of relative
atomic size were derived from Lothar Meyer's table of atomic
volumes).

PROBLEM: Look up the decomposition temperatures of the carbonates
in Main Groups 1 & 2 and attempt to explain the trends.

A.3
The empirical shape of the modern Periodic Table

Modern arrangements of the periodic table differ considerably from that of Mendeleev. The five principal refinements can all be justified empirically, but their general acceptance has been hastened because they reflect the enormous advances in theoretical physics which have taken place. As always, science advances when there is a fruitful interplay between hypothesis and experiment.

A.3(1) SEPARATION OF MAIN GROUPS FROM TRANSITIONAL GROUPS

The fourth Period was arranged by Mendeleev in the following way:

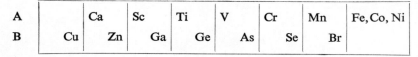

								Transitional
A		Ca	Sc	Ti	V	Cr	Mn	Fe, Co, Ni
B	Cu	Zn	Ga	Ge	As	Se	Br	

However, the differences between the A and B elements in the same Group are in most cases very much greater than their similarities as the following examples show: (see p. 56 for a demonstration of the gross dissimilarities between potassium and copper).

Group 5: Vanadium is a hard high melting point metal, forming a non-stoichiometric hydride. It forms no less than four series of coloured salts, in which the oxidation state of vanadium varies from $+2$ to $+5$. Arsenic is a metalloid with a volatile hydride AsH_3. It forms two series of salts which are not fully ionised and extensively hydrolysed in solution.

The highest oxides of both elements are acidic, and react with alkalis to form arsenates and vanadates which are isomorphous. These compounds will oxidise I^- to I_2, and are reduced by zinc and dilute sulphuric acid.

Group 7: Manganese is a hard high melting point metal yielding hydrogen with acids with a non-stoichiometric insoluble hydride. It forms stable salts in its $+2$ oxidation state, and all other oxidation states up to $+7$ are known. The black non-stoichiometric oxide MnO_2 decomposes on heating to give Mn_3O_4; it will dissolve both in concentrated acids and alkalis. Chlorine is a gaseous non-metal with a volatile hydride HCl which dissolves in water to give an acid solution. Chlorine forms stable salts in which the Cl has an oxidation number of -1. The gaseous oxide ClO_2 is violently explosive, and dissolves only in alkalis.

Both elements form an unstable volatile oxide X_2O_7 which dissolves in alkalis to form salts which give oxygen on heating.

The only case where there are considerable similarities between the A Group and B Group elements is Mendeleev's Group 2, as the following comparison between calcium and zinc shows.

Group 2: Both calcium and zinc are silvery metals of fairly high melting point. Both burn in air to form a white oxide which is stable to heat; this oxide dissolves in acids to form colourless salts containing a M^{2+} ion; $+2$ is the only oxidation state shown by both elements. Both elements combine directly with the halogens and sulphur; the halide salts are soluble and the sulphides of both elements are decomposed by acids.

However, there are certain differences. Calcium liberates hydrogen from water, which zinc does not. Both hydroxides are sparingly soluble, but that of zinc can be dehydrated by heat and dissolves in alkalis to give zincates. Calcium hydride is salt-like, whereas that of zinc is ill defined and cannot be made by direct combination. Zinc forms many alkyls, whereas calcium does not. Zinc salts are more hydrated than calcium salts and dissolve to give acid solutions.

Altogether the properties of zinc are similar to those of an imaginary element lying between magnesium and beryllium in M2. A similarity as close as this should certainly be reflected in an empirical periodic table.

A.3(2) THE CONTINUITY IN PROPERTIES BETWEEN Ti, V, Cr, Mn AND Fe, Co, Ni

All these elements have many points in common:
(1) They are all high melting point metals.
(2) They do not react with water.
(3) All these elements have coloured insoluble oxides.
(4) All of them form more than one series of coloured ions.

There is no case therefore for classifying these elements separately as A Group elements and 'transitional' elements as Mendeleev did.

The final blow to Mendeleev's octave arrangement, with properties repeating at the eighth element with iron, cobalt and nickel as an arbitrary interlude, was the discovery of the noble gases in 1895; for these involved yet another transition between A Groups and B Groups.

The following arrangement is the modern solution: T3–T11 occur as a separate transition series between M2 and M2'.

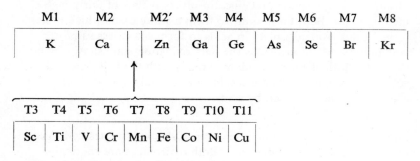

M1	M2		M2'	M3	M4	M5	M6	M7	M8
K	Ca		Zn	Ga	Ge	As	Se	Br	Kr

T3	T4	T5	T6	T7	T8	T9	T10	T11
Sc	Ti	V	Cr	Mn	Fe	Co	Ni	Cu

This arrangement has the following merits:
(1) It stresses the similarity between Ca and Zn.
(2) It stresses the fundamental difference between K and Cu.
(3) It stresses the similarity between Ti, V, Cr, Mn and Fe, Co, Ni.
(4) It indicates a faint similarity between the former A Group and B Group elements by referring to them by the same Group number; e.g. vanadium is in T5, whereas arsenic is in M5.

A.3(3) THE INNER TRANSITIONAL ELEMENTS

The insertion of the transitional elements into the modern empirical table as a separate system below the Main Groups is very satisfactory.

A similar treatment can also be used to accommodate the two series of inner transitional elements as a separate system below the transitional elements.

Such an arrangement brings out the very great similarities of the elements to one another. The older term 'rare earth' is a poor name for the first series, since they can be obtained quite readily as one of the products of uranium fission—they are better called 'lanthanons'. ✓

(1) They all (except La and Lu) exhibit paramagnetism.
(2) They are all electropositive metals.
(3) They all have an oxidation number of $+3$ which is very stable.
(4) Their salts tend to be isomorphous and difficult to separate.

Other oxidation states are rare and unstable with respect to the oxidation state of $+3$: for instance the orange cerium (IV) is a volumetric oxidising agent, readily reduced by acidified Fe^{2+}.

A.3(4) DISCONTINUITY IN THE LATER MAIN GROUPS

The later Main Group elements show a discontinuity between Period 3 and Period 4. A particularly notable feature of this discontinuity is the relative instability of the high oxidation states of the Period 4 elements.

M5. PCl_5 and $SbCl_5$ dissociate on heating, but $AsCl_5$ does not exist; similarly PF_5 and SbF_5 are very much more stable than AsF_5.

M6. SeO_3 dissociates on heating much more easily than SO_3 or TeO_3.

M7.   Perchlorates and periodates of the general formula XO_4^- are easily prepared and so are the corresponding acids. Perbromates and perbromic acid have been reported, but they are very unstable.

There are many other examples which will be dealt with in the chemistry of the relevant Groups. This anomalous discontinuity is best brought out by leaving a small gap between Period 3 elements and Period 4 elements in the later Main Groups.

A.3(5) THE UNIQUE POSITION OF HYDROGEN

Hydrogen has a genuine claim to a place in any of the Groups M1, M4, M7; as the following experiments show.

M1.　(a) Hydrogen will, under enormous pressure, adopt a metallic structure.
(b) Its compounds with the halogens, when dissolved in water, will give rise to hydrated mono-positive ions.
However, its compounds with the halogens are gaseous at room temperature, not high melting crystalline solids; nor will the alkali metals ever form negative ions, as hydrogen will.

M7.　(a) Like the halogens, hydrogen is a very volatile diatomic gas at room temperature.
(b) Hydrogen will exist as negative ions in its compounds with the alkali metals. These compounds are salt-like and conduct electricity when molten.
(c) Hydrogen will form complex ions, BH_4^-, AlH_4^- which are similar to halide complexes like BF_4^-, $AlCl_4^-$.
However, its compounds with the alkali metals, unlike the halides, will decompose immediately with water. Also its oxide is stable, which the oxides of the halogens certainly are not.

M4.　(a) Like CO, the oxide of hydrogen is volatile and thermally stable.
(b) Compounds of both hydrogen and carbon with the halogens are volatile and thermally stable.
However, carbon forms very unstable ions whereas hydrogen forms stable cations and anions. Again, the halides of carbon are relatively inert, whereas those of hydrogen are very reactive.

So although hydrogen has certain claims to inclusion in no less than three Main Groups, it has in each case contradictory properties which ensure that any one classification is unrealistic. Accordingly it is placed in a unique position in the table.

The Periodic Table (Figure 2) which will be used throughout the remainder of the book contains all these improvements. Its form was first suggested by Professor R. T. Sanderson of Iowa University in 1964 (JCE 1964, page 187).

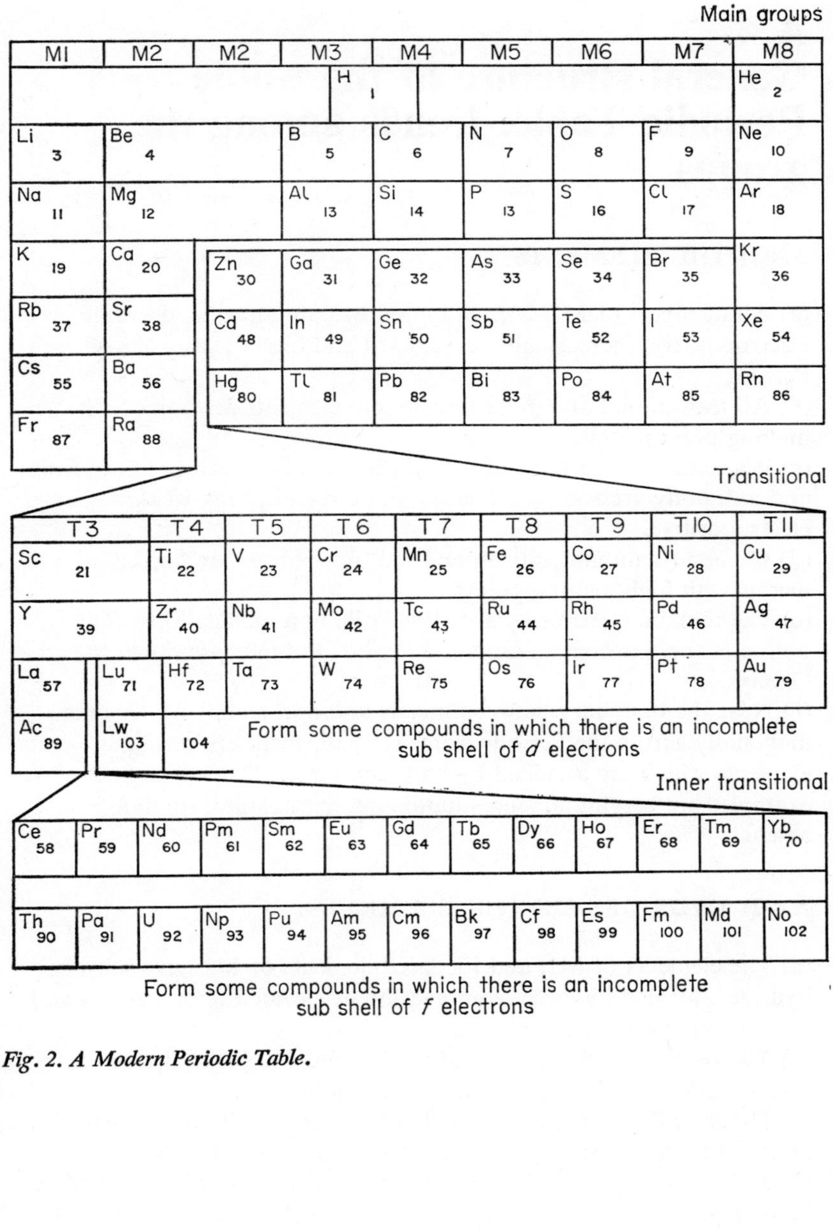

Main groups

M1	M2	M2	M3	M4	M5	M6	M7	M8
			H 1					He 2
Li 3	Be 4		B 5	C 6	N 7	O 8	F 9	Ne 10
Na 11	Mg 12		Al 13	Si 14	P 15	S 16	Cl 17	Ar 18
K 19	Ca 20	Zn 30	Ga 31	Ge 32	As 33	Se 34	Br 35	Kr 36
Rb 37	Sr 38	Cd 48	In 49	Sn 50	Sb 51	Te 52	I 53	Xe 54
Cs 55	Ba 56	Hg 80	Tl 81	Pb 82	Bi 83	Po 84	At 85	Rn 86
Fr 87	Ra 88							

Transitional

T3	T4	T5	T6	T7	T8	T9	T10	T11
Sc 21	Ti 22	V 23	Cr 24	Mn 25	Fe 26	Co 27	Ni 28	Cu 29
Y 39	Zr 40	Nb 41	Mo 42	Tc 43	Ru 44	Rh 45	Pd 46	Ag 47
La 57	Lu 71	Hf 72	Ta 73	W 74	Re 75	Os 76	Ir 77	Pt 78	Au 79
Ac 89	Lw 103	104						

Form some compounds in which there is an incomplete sub shell of d electrons

Inner transitional

Ce 58	Pr 59	Nd 60	Pm 61	Sm 62	Eu 63	Gd 64	Tb 65	Dy 66	Ho 67	Er 68	Tm 69	Yb 70
Th 90	Pa 91	U 92	Np 93	Pu 94	Am 95	Cm 96	Bk 97	Cf 98	Es 99	Fm 100	Md 101	No 102

Form some compounds in which there is an incomplete sub shell of f electrons

Fig. 2. A Modern Periodic Table.

A.4
General structure of the whole Periodic Table: trends among the groups

A.4(1) THE ELEMENTS

(a) All members of M1, M2, M2', and M3 are metallic, decreasingly electropositive, increasingly dense, hard and high melting across each Period.

(b) All transitional and inner transitional elements are hard, high melting point metals.

(c) White tin and lead, bismuth and the metallic allotropes of arsenic and antimony are soft, low melting point metals, only weakly electropositive.

(d) Carbon (diamond), silicon and red phosphorus are hard non-metals with high melting points.

(e) The remaining elements are all volatile non-metals. The boiling points of these decrease across the Periods but increase down the Groups.

(f) Near the boundary between metals and non-metals which runs diagonally across the Periods from Al to Bi, there are certain elements which are attacked by both acids and alkalis. The best examples are beryllium, zinc, aluminium, germanium, tin and arsenic.

A.4(2) HYDRIDES OF THE ELEMENTS

(a) The elements of M1, and the later elements of M2 have salt-like hydrides, formed by direct combination at elevated temperatures and decomposing on strong heating.

(b) The remaining elements of M2, M2', and M3 form ill defined non-volatile hydrides which decompose on heating.

(c) The hydrides of the transitional and inner transitional elements are hard high melting non-stoichiometric compounds.

(d) The remaining hydrides are all gases at room temperature or low boiling liquids. The hydrides of boron, the M4 elements and the M5 elements are unstable to heat and readily oxidised. Hydrides of M6 and M7 elements show acidic properties when dissolved in water. The boiling points of the hydrides of nitrogen, oxygen and fluorine do not fit in with an extrapolation of the Group trend.

A.4(3) OXIDES OF THE ELEMENTS

(a) The trend across the Periods is from highly stable crystalline solids of high melting point which dissolve in water to give alkaline solutions, through solids which are insoluble and infusible, to increasingly volatile and unstable molecular solids which dissolve in water to give acidic solutions. (This trend holds for the oxides of the elements in their Group oxidation state; there are of course numerous lower oxides of the later Main Groups, from M4 onwards. Higher oxides can be reduced to lower oxides, and sometimes to the element, by heating with carbon; indeed some higher oxides decompose spontaneously on heating, especially in M6 and M7 where they are often strong oxidising agents.)

(b) Oxides of the early Main Group elements (basic oxides) will readily react with oxides of the later Main Group elements (acidic oxides) forming complex oxides, better known as salts. These are ionic compounds and dissolve in water.

$$Li_2O + SO_3 \rightarrow Li_2SO_4$$

(c) Besides being a splendid solvent for salts, the oxide of hydrogen has amphoteric properties, which means that it can form conducting solutions with both basic and acidic oxides:

$$Li_2O + H_2O \rightarrow 2LiOH\text{—a strong alkali}$$
$$SO_3 + H_2O \rightarrow H_2SO_4\text{—a strong acid}$$

For an acid with the general formula $XO_m(OH)_n$, the higher the value of m, the stronger the acid.

The infusible and insoluble oxides of elements towards the middle of the Periods, for convenience called 'intermediate oxides', often show amphoteric properties by reacting with both acidic and basic oxides; the resulting solutions conduct electricity when dissolved in water. Amphoteric oxides themselves react only to a very limited extent with water, since oxides of the same class do not generally combine with each other.

(d) Though they do not combine with water or dissolve in it, amphoteric oxides will dissolve in solutions containing an excess of H_3O^+ or an excess of OH^-.

There are some intermediate oxides which are not amphoteric. These in the early Main Groups tend to dissolve in acids

$$MgO + 2H_3O^+ \rightarrow Mg^{2+} + 3H_2O$$

Those in the later Main Groups tend to dissolve in alkalis

$$SiO_2 + 2OH^- \rightarrow SiO_3{}^{2-} + H_2O$$

(e) In the transitional Periods, oxides of the lower oxidation states dissolve in acids, e.g. MnO, while oxides of the higher oxidation states dissolve in alkali, e.g. CrO_3. The latter solutions reprecipitate the oxide when strong acid is added.

(f) Oxyanion stability depends upon the nature of the cation. Hydrated salts are more stable than anhydrous salts of the same cation, which in turn are more stable than anhydrous acids. Thus hydrated copper nitrate crystals are stable, whereas anhydrous copper nitrate is not, and must be prepared by a special method using liquid N_2O_4. $KClO_4$ is stable at above 300°C, whereas $HClO_4$ is dangerously explosive when heated.

A.4(4) HALIDES OF THE ELEMENTS

(a) The trend across the Periods is from thermally stable high melting point salt-like compounds soluble in water to relatively unstable volatile substances which react with water.

(b) The halides of the transitional metals, like those of M3, are almost always strongly hydrated and extensively hydrolysed in solution.

(c) The higher the Main Group number of a particular element, the more likely that its halide will:

   (i) act as a halogenating agent by giving up halogen to other molecules;

   (ii) hydrolyse in water with the formation of hydrohalic acid.

(d) The relative volatility of the halides changes with the Main Group number of the element. For the metals of M1 and M2, (which give basic oxides) the fluoride tends to be the least volatile and the iodide the most volatile of the salts. For the molecular halides of the non-metals (which give acidic oxides) the order of volatility is reversed; thus BF_3 boils at -100°C and BI_3 at 49°C. For the halides of elements intermediate between M2 and non-metals the fluoride tends to be very involatile, whereas the volatility of the other halides is low but increases from chloride to bromide to iodide.

A.5
Trends among Periodic Groups

A.5(1) GROUP M1: THE ALKALI METALS

Lithium

SODIUM
POTASSIUM
Rubidium
Caesium

Apart from the usual anomalous properties of the first member of the Group, the remaining M1 elements are very similar indeed. They are increasingly soft silvery white metals, tarnishing rapidly in air; they are very electropositive, so they are prepared by the electrolysis of their fused chlorides. The melting point decreases from Li (186°C) to Cs (28°C), and the density rises from 0·53 g cm^{-3} to 1·87 g cm^{-3}; sodium will float on water whereas caesium will not. Lothar Meyer's curve of the atomic volumes of elements shows local maxima for M1 metals (p. 5) and they are all excellent conductors of electricity. They form amalgams with mercury, and can be dispersed in hydrocarbons and liquid ammonia: the latter solution is unstable and decomposes with a trace of catalyst to sodamide and hydrogen. Just above the boiling points, about 1% of the atoms exist as diatomic molecules: Li_2 has the highest heat of dissociation.

The reactivity of the M1 metals to all chemical reagents except nitrogen is very high and increases down the group. Li is only slowly attacked by water, Na reacts vigorously and K inflames. All the elements burn in air; Li forms Li_2O, Na forms Na_2O_2 and potassium KO_2. Lithium reacts much less readily with hydrogen (750°C) than Na (400°C) but the compound so formed is much more stable than the hydrides of the other M1 metals, and can be melted without decomposition at 668°C. These hydrides are insoluble in organic solvents and are decomposed explosively by water. The metals will replace the acidic hydrogen in hydrocarbons such as acetylene and cyclopentadiene, lithium less readily than the others: in contrast to the salt-like sodium alkyls, which do not dissolve in hydrocarbons, lithium alkyls [except the polymeric $(LiCH_3)_n$] are low melting solids and are miscible with organic solvents. Direct reaction of the metals with other elements can lead to borides, silicides, arsenides and sulphides; these are salt-like and are readily hydrolysed by water and dilute acids.

The oxides dissolve readily in water to give strongly alkaline solutions; peroxides and superoxides liberate oxygen in addition. The hydroxides so formed are all soluble and their solutions are completely ionised. Lithium hydroxide is less soluble than the others, and is the only one to be decomposed at red heat.

The M1 metals form only one series of salts—which are always colourless unless the anion is coloured—and practically no complexes. The salts of strong acids are almost all freely soluble (exceptions are the perchlorates and hexanitrocobaltates(III) of K, Rb and Cs) and increasingly thermally stable; solid hydrogen carbonates are formed only by M1 metals. Salts of strong acids melt before they decompose, whereas the salts of weak acids nearly all decompose before they melt. Once again lithium is anomalous, since its carbonate decomposes at 600°C, and this salt, the phosphate and the fluoride are insoluble (cf Mg, p. 51). The majority of lithium and sodium salts contain water of crystallisation; about a quarter of the salts of potassium are hydrated, but those of Rb and Cs very seldom are. Lithium ions are also very strongly hydrated in solution, and it is not surprising that the ionic mobility of the M1 ions increases down the group.

All their volatile compounds produce flame spectra characteristic of the metal they contain: the sodium flame is yellow, that of potassium is a combination of blue and rose.

A.5(2a) GROUP M2: THE ALKALINE EARTHS

Beryllium
_____

MAGNESIUM
_____

CALCIUM
STRONTIUM
BARIUM
Radium

The elements of M2, from Mg to Ba, bear a close resemblance to the elements of M1, from Li to Cs; beryllium, the first member of M2, shows unique chemical properties which will be dealt with at the end of this section.

The M2 metals are rather harder and more dense (Mg 1·74 g cm^{-3} Ba 3·78 g cm^{-3}) than the equivalent M1 elements, and are also formed by the electrolysis of the fused chlorides owing to their highly electropositive character. The silvery white metals are excellent conductors of electricity; all of them readily amalgamate with mercury, even though magnesium is protected by a grey layer of metallic oxide (cf. aluminium).

The chemical reactivity of the M2 metals is high and generally increases down the Group. Thus Mg will only react with steam at 200°C, whereas Ba will react rapidly with water at 25°C. All the elements burn in air to the oxide MO, barium alone BaO_2. The hydrides are formed increasingly easily by direct combination; that of magnesium is a grey polymeric powder, whereas the others are salt-like, yielding hydrogen readily with water and acids. Magnesium, like lithium, readily forms a nitride stable to heat but hydrolysed by water to give ammonia. The metals combine with carbon in an electric furnace at over 1500°C to yield salt-like acetylides with the general formula MC_2, which hydrolyse readily with water and acids to form acetylene. Similarly borides, nitrides, arsenides and sulphides are all made by direct combination, and being salt-like in nature undergo a similar hydrolysis. The salt-like alkyls of Ca, Sr and Ba are unimportant, but magnesium alkyls (Grignard reagents) are low boiling liquids and solids, easily formed by the action of the relevant alkyl or aryl halide in ethereal solution on the metal.

Magnesium oxide is almost insoluble, but the solubility increases down the Group and barium oxide will dissolve to give a fully ionised carbonate-free solution M/10 with respect to OH$^-$. The M2 metals form only one series of salts which are colourless unless the anion is coloured. The salts of strong acids are decreasingly soluble down the Group ($MgSO_4 \cdot 7H_2O$ freely soluble, $BaSO_4$ insoluble) and increasingly thermally stable, though not quite as stable as the equivalent compounds of M1. Once again salts of the strong acids tend to melt before they decompose, while salts of weak acids always decompose before they melt. Salts of weak acids tend to be sparingly soluble (often increasing irregularly down the Group) in water, but soluble in dilute solutions of strong acids. The solubility of the fluorides also increases down the Group. Magnesium salts almost always contain water of crystallisation, calcium salts very often do, while barium salts only occasionally contain it. $BaCl_2.2H_2O$ is easily dehydrated at 300°C, whereas $MgCl_2.6H_2O$ forms $Mg(OH)Cl$ when heated strongly. All the volatile compounds of M2 elements except Mg give visible flame spectra: the latter emits strongly in the near ultraviolet.

All the M2 metallic ions form stable complexes with EDTA (used for quantitative estimation) and polyphosphates (used for softening water), but other complexes of Ca, Sr and Ba are much less easily formed than those of Mg. Grignard reagents form addition compounds with oxygen-containing organic compounds, and $Mg(ClO_4)_2$ will dissolve readily in alcohols and ethers. The chemistry of magnesium is similar to that of lithium (see p. 149) in its tendency to form complexes.

Beryllium is a metal, but the chemistry of this element can scarcely be called metallic. The simple Be^{2+} ion is never formed even in BeF_2, and $Be(H_2O)_4{}^{2+}$ is only stable in concentrated acids. There are a number of basic salts, e.g. $BeO : 3Be (acetate)_2$, in which the beryllium has formed a strong bond with the oxide anion. Not surprisingly therefore, beryllium hydroxide dissolves in both acids and alkalis, and beryllates are well-defined compounds. Beryllium chloride hardly conducts electricity when fused; it melts at 450°C, far lower than the other M2 halides and the molecules are dimeric in the vapour phase. Beryllium hydride is thermally unstable and polymeric, and beryllium forms a number of low melting alkyls. The chemistry of beryllium is similar to that of aluminium (see p. 149).

A.5(2b) METALS IN THE LATER MAIN GROUPS

The properties of metals from M2′ onwards demonstrate the unsatisfactory nature of all hard and fast classification; to define a metal as an element which readily conducts electricity focuses attention on one characteristic physical property of limited importance, and tends to obscure the *gradual* transition in chemical properties which takes place across the Periods and down the later Main Groups. Lead and sodium are different kinds of elements which both happen to conduct electricity when they are pure—so they both appear under the blanket heading 'Metals'; in chemical properties they certainly differ more from each other than lead and silicon, yet the latter are classified under the separate headings of 'Metal' and 'Non-metal'.

The chemical properties of M1 and M2 which are characteristic of these elements can be summed up as follows.

Strongly electropositive, reacting readily with acids.

In their Group oxidation state they form basic oxides which are readily soluble even in weak acids.

Form many stable salts with oxyacids, which are not hydrolysed in solution.

Form comparatively few complexes.

React exothermically with non-metals.

The metals from M2' onwards across the Periods fit these criteria increasingly less well. The metals of M2' and to a lesser extent M3 form cations in the Group oxidation state which are still recognisably like those of the metals in M1 and M2; the similarity of calcium to zinc is unmistakable if not exact (p. 57). By M4, however, the Group oxidation state is almost exclusively acidic; Sn^{4+} and Pb^{4+} do not exist, the hydrides and chlorides are volatile and the higher oxides dissolve in alkali. The same is true of Sb and Bi, which are low melting and volatile metals—the furthest from the left of the Table to conduct electricity. The Group oxidation state shows a regular increase in oxidising power across the Periods, especially in Period 6; it is somewhat more stable under alkaline conditions than in acidic solutions.

All the metals in the later Main Groups form some compounds in which they show a stable oxidation state of two less than their Group oxidation number. Mercury does so in the exceptional $[Hg(0)—Hg(II)]^{2+}$ ion, TlOH is an alkali, Pb^{2+} is well defined, and Sn^{2+} and Bi^{3+} exist despite their strong tendency to hydrolysis. The dioxides of the M6 elements and the XO_3^- compounds in M7 complete the pattern. In the later Main Groups there are oxidation states even further below the Group number: there are some traces oı cationic behaviour for Po(II) and the cations I^+ and I^{3+} are both known.

The later Main Group metals occur in nature as their sulphides, which are brightly coloured and very insoluble, even in acid solution. The sulphides can always be converted to the oxides by roasting them in air at high temperatures.

Ions in the lower oxidation states of later Main Group metals are only formally similar to the ions of M1 and M2. Very often the salts crystallise with non-uniform lattices in which similar atoms do not all have the same environment. The hydroxides are usually amphoteric, and many basic salts are formed. The iodides tend to be the most insoluble of the halide salts and it is iodides rather than fluorides which form the most stable anionic complexes. Sulphide complexes are stable, and preferred when there is a choice, as in the SCN^- ion which can form complexes using either the N or the S atom; similarly the decomposition of thiosulphates leads to the sulphide rather than the oxide of the metal. Ammines similar to those formed by ions of transitional metals, e.g. $Cu(NH_3)_4^{2+}$, are either relatively unimportant or not formed at all; nor are complexes with unsaturated ligands like CO, CN^- or ethylene.

In the later Main Groups, there is no doubt that the difference between, say, Si and Pb is greater than that between Na and Cs; but it is possible to over-emphasise the contrast between the former pair by harping on the all-or-nothing distinction between metals and non-metals. There are in fact very close similarities between Si and Pb in the Group oxidation state which emphasise the non-metallic chemistry of the latter. Both are resistant to attack by dilute acids and alkalis but dissolve in concentrated HNO_3. Their dioxides are soluble in alkali. They form volatile tetrachlorides which are readily hydrolysed. The tetrahydrides are unstable and pyrophoric, and the tetraethyls are low boiling liquids. A comparison between P and Bi in their Group oxidation state will reveal that they also are very similar. It is only in the lower oxidation states that the later Main Group elements reveal their ambiguity and masquerade as metals.

A.5(2c) GROUP M2′

ZINC
Cadmium
MERCURY

The elements of M2′ are all comparatively volatile metals of low melting point. They are much denser than their analogues in M2 and much less electropositive: unlike M2 the electropositivity decreases down the Group, and mercury is only attacked by acids that are strong oxidising agents.

Oxides are formed by direct combination when the metals are heated in air; they are coloured under certain conditions (ZnO is yellow when hot but white when cold). ZnO and CdO are stable to heat but HgO is decomposed at 500°C. Zinc hydroxide is amphoteric, cadmium hydroxide less so while mercury hydroxide (which scarcely exists, turning rapidly into HgO) has no tendency whatever to dissolve in alkalis.

ZnS is soluble in dilute HCl (but not dilute acetic acid), CdS is soluble in concentrated acids whereas HgS is insoluble in all acids unless they are strong oxidising agents.

Halides are prepared by direct action. The fluorides melt at high temperatures; they are rather insoluble owing to their high lattice energy. HgF_2, the salt of a weak acid and a very weak base, is extensively hydrolysed in solution. Other halides are less salt-like, soluble in organic solvent and low melting. They are soluble in water also forming a multitude of complex ions: $HgCl_2$ dissolves to a limited extent as unionised molecules, but the scarlet HgI_2 is very insoluble.

The pure metals do not dissolve in strong acids to form salts and hydrogen: Zn and Cd do so if they contain metallic impurities which modify surface properties. However, they dissolve freely under oxidising conditions to give strongly hydrated nitrates ahd sulphates which hydrolyse in solution unless excess acid is present. Salts of weak acids are frequently basic salts and sparingly soluble; they are prepared by double decomposition.

There are an enormous number of complexes of M2$'$ elements, but these are not as stable as the complexes of the transitional metals. Examples are $Zn(H_2O)_4{}^{2+}$, $CdCl^+$, $Cd(CN)_4{}^{2-}$, $Zn(NH_3)_4{}^{2+}$—the latter accounts for the immediate solubility of $Zn(OH)_2$ in the NH_4OH/NH_4Cl mixture used in qualitative analysis. Mercury forms complexes through almost any atom except oxygen: the action of ammonia on mercury(II) chloride solution gives three different complexes depending upon the condition used, and excess I^- converts the insoluble scarlet HgI_2 to colourless $HgI_4{}^{2-}$ which dissolves.

There are multitudinous alkyls of mercury, made by adding anhydrous $HgCl_2$ to the appropriate Grignard reagent:

$$HgCl_2 + 2RMgI \rightarrow R—Hg—R + 2MgICl$$

They are toxic low melting compounds stable to air and water, so they can be stored. Alkyls of mercury can be used to prepare the alkyls of many other elements, Zn and Cd among them:

$$HgR_2 + Zn \rightarrow ZnR_2 + Hg$$

Such compounds are readily attacked by atmospheric oxygen and violently hydrolysed by water.

Mercury is unusual in forming a second series of salts based on the Hg_2^{2+} ion. The mercury(I) salts that exist are rather similar to the equivalent salts of silver: thus mercury(I) nitrate (made by reacting mercury with hot mercury(II) nitrate solution) is soluble, and mercury(I) chloride is formed as a white precipitate immediately chloride ions are added to the solution. Mercury(I) salts can never be formed when the equivalent mercury(II) salt is rather insoluble, and any attempt to form mercury(I) complexes always leads to the formation of the corresponding mercury(II) complex and the deposition of metallic mercury. Strong reducing agents, like $SnCl_2$ and formate ions, reduce mercury(II) compounds to mercury(I) compounds, but excess of the reagent leads to further reduction and grey metallic mercury is deposited.

A.5(3) ALUMINIUM AND THE ELEMENTS OF M3

Boron

ALUMINIUM
Gallium
Indium
Thallium

Aluminium and the succeeding elements in M3 are clearly metallic, but their compounds are by no means genuinely salt-like and many complexes are formed. Since it is the commonest metal in the crust of the earth, the chemistry of aluminium is exceedingly important; the remaining elements in M3 are rare. The Group trends tend to be ill-defined: thus there is a discontinuity between the high electropositivity of aluminium and the low and decreasing electropositivity of the remaining elements (there is a parallel, however, in the series Mg, Zn, Cd, Hg). The salts of strong acids are strongly hydrated and extensively hydrolysed: the oxides and hydroxides change from being amphoteric to basic down the Group. The tendency to complex formation increases down the Group, and so does the importance of an oxidation state of $+1$ (the 'inert pair' effect, p. 123).

Elemental boron is very difficult to prepare in a pure state. Its chemistry is exceptional, bearing some resemblances to that of silicon (p. 52).

Aluminium is made by the electrolysis of its purified oxide in a bath of molten Na_3AlF_6. It is normally dull grey in colour owing to a tough surface layer of aluminium oxide; when freshly scratched however it is a bright silver colour. The oxide layer is often thickened artificially by anodic oxidation and in this state it will absorb pigment into the pores of the oxide layer.

Aluminium will not be attacked by water or steam unless it is first amalgamated. It will however dissolve readily in dilute acids unless it is very pure, and it gives hydrogen with solutions of warm alkali, especially if alloyed with copper. Hot concentrated nitric acid renders it passive. It reacts directly with halogens, nitrogen, carbon, and other non-metals except hydrogen; an ill-defined solid hydride, probably polymeric, has been reported.

Aluminium oxide is amphoteric, dissolving in acids to form $Al(H_2O)_6{}^{3+}$ and alkalis to form $Al(OH)_4{}^-$. Even the salts of strong acids are extensively hydrolysed in solution according to the equation:

$$Al(H_2O)_6{}^{3+} + H_2O \rightarrow Al(H_2O)_5OH^{2+} + H_3O^+.$$

The resulting solutions are so acidic that any attempt to make the salts of weak acids—aluminium carbonate or sulphide for example—by double decomposition will only lead to a solution of the free weak acid and a precipitate of aluminium hydroxide.

The anhydrous chloride is made by direct combination: it sublimes at 185°C in the form of the dimer Al_2Cl_6 (p. 125). It is used extensively as a catalyst in organic chemistry—it forms $R^+AlCl_4{}^-$ with alkyl chlorides and R_3N—$AlCl_3$ with tertiary amines. Aluminium forms a large range of pyrophoric alkyls; those containing a small number of carbon atoms are dimeric.

Both tetrahedral ($AlCl_4{}^-$, $AlH_4{}^-$) and octahedral ($Al(H_2O)_6{}^{3+}$, $AlF_6{}^{3-}$) complexes are formed, but no compounds containing ammonia (e.g. $Al(NH_3)_6{}^{3+}$) are stable because hydrolysis takes place

$$Al(NH_3)_6{}^{3+} + H_3O^+ \rightarrow Al(NH_3)_5H_2O^{3+} + NH_4{}^+.$$

The trend across a complete Period from metal to non-metal is

faithfully reflected in the gradation of properties from
Na $\rightarrow$ Mg $\rightarrow$ Al, the first three elements in Period 3.

Elemental boron is a very hard high melting crystalline substance
which does not conduct electricity. It is very inert to chemical attack,
unaffected by dilute acids and only dissolving in fused alkalis or hot
concentrated nitric acid. The chemistry of boron is entirely non-
metallic; the B^{3+} ion is never found.

Compounds of boron with elements less electronegative than itself
are called borides. They have surprising formulae, e.g. CaB_6, and
are prepared by direct combination at high temperatures. Magnesium
boride readily reacts with water to give mixtures of boron hydrides.
These compounds are mostly pyrophoric and violently hydrolysed.
Salt-like lithium borohydride, $Li^+BH_4^-$ is formed by the action of
lithium hydride and B_2H_6 in ether. It is a strong reducing agent.

The halides are made by direct combination; they are monomeric
and increasingly easily hydrolysed. BF_3 very readily forms addition
compounds with water, alcohols, ethers, amines and phosphines, and
forms the complex ion BF_4^- with fluoride ions. The other boron
halides behave in the same way but the bonds formed are not so
strong.
    Boric acid is formed by the hydrolysis of the hydrides and
chlorides; it will dehydrate on heating. It is soluble in water and
weakly acidic owing to the ionisation

$$B(OH)_3 + 2H_2O \rightarrow H_3O^+ + B(OH)_4^-$$

An extensive chemistry of condensed and polymeric borates exists.

There is a vast organic chemistry of boron also, based on boron
alkyls, esters of boric acid and compounds containing both types of
linkage.

By substituting alternate B and N atoms for successive carbon atoms
some pseudo-organic compounds can be synthesised. $(BN)_x$ is almost
as hard as diamond, and $B_3N_3H_6$ resembles benzene in some of its
physical and chemical properties.

A.5(4) GROUP M4

CARBON

SILICON
Germanium
TIN
LEAD

The change from a diamond-like three dimensional structure to a
true metallic crystal takes place at tin; the grey allotrope stable at
low temperatures is non-metallic whereas white tin behaves as a true
metal. However, an arbitrary division between metals and non-metals
in the Group is unhelpful, for it misrepresents the gradual nature of
the chemical trends. The unique chemistry of carbon is considered
separately.

Silicon is a very inert element, but there is a small increase in
chemical reactivity down the group (lead often forms an
impenetrable surface layer of compound with many reagents which
inhibits further change). Air and water, dilute acids and alkalis all
only react at red heat. Unlike carbon, silicon reacts with chlorine,
and the other elements do so increasingly easily; the tetrahalides so
formed (lead gives mostly $PbCl_2$) are low boiling liquids. The only
acid which can attack silicon is HF, forming H_2SiF_6, but the
remaining elements will also dissolve in concentrated HNO_3 in
typical non-metallic fashion forming the dioxides (except in the case
of lead, which forms $Pb(NO_3)_2$).

The tendency to catenation (chain formation) which dominates the
chemistry of carbon is also found in the hydrides of silicon and
germanium, but no hydrides containing more than one metal atom
have been found for tin and lead. The temperature at which the XH_4
compound decomposes decreases down the Group. Germanium
shows some exceptional properties—its maximum chain length in
Ge_8H_{18} is longer than that of the highest silane, germanium hydrides
are far more resistant to alkaline hydrolysis than the silanes, they do
not inflame so easily, and, like AsH_3, GeH_4 can be prepared by
reduction of germanium halides, using Zn/HCl.

The halides of silicon including SiF_4 react violently with water,
whereas those of the remaining elements are reversibly hydrolysed
only. SiF_4 and GeF_4 are gases at room temperature, but SnF_4 and
PbF_4 sublime at high temperatures. The formation of ions such as

XF_6^{2-} and XCl_6^{2-} is very common even for lead. $PbBr_4$ and PbI_4 do not exist because of the strong oxidising properties of lead(IV).

The dioxides show a well-defined trend: SiO_2 is exclusively acidic, GeO_2 will just dissolve in concentrated HCl, whereas SnO_2 is amphoteric. PbO_2 is so insoluble that it is difficult to classify; it is a very strong oxidising agent which is used in the lead accumulator. There is an enormous silicate chemistry of commercial importance, since the Si–O bond is so very stable and silicon is such a common element. There are a few orthosilicates containing discrete SiO_4^{4-} ions, each oxygen atom being bonded to a divalent metal ion to form a three dimensional structure of high melting point. In addition a vast number of chain, cyclic, sheetlike and framework silicates are formed by condensation and polymerisation. Both orthogermanates, based on the GeO_4^{4-} ion, and metagermanates are known, while the larger tin and lead atoms form the octahedral $X(OH)_6^{2-}$ ion.

A certain number of silicides, and to a lesser extent germanides, are known. When combined with the electropositive metals compounds are formed which hydrolyse with acids to give the M4 hydride, whereas the transitional metals form a range of solid solutions and non-stoichiometric compounds with the M4 elements.

SnS_2 is precipitated from tin(IV) solutions with H_2S; it is insoluble in acids but dissolves in ammonium sulphide solution to give thiostannates. Lead salts precipitate PbS which is also insoluble in acids.

There is an extensive organic chemistry of the M4 elements. The tetraethyls are increasingly unstable, yet large quantities of $Pb(C_2H_5)_4$ are manufactured for use as an anti-knock. The silicones, formed by the hydrolysis of R_2SiCl_2 followed by polymerisation, with

$$-O-\overset{\displaystyle R}{\underset{\displaystyle R}{\overset{|}{\underset{|}{Si}}}}-O-$$

as the basic unit, are very important as water-repellents. R_3SnCl is interesting in that it ionises to $R_3SnOH_2^+Cl^-$.

The oxidation state of $+2$ is very unstable for the elements silicon and germanium; Sn(II) shows its relative instability w.r.t. Sn(IV) by its strong reducing powers in both acid and alkaline solution, while Pb(II) is far more stable than Pb(IV), especially in acid solution.

Anhydrous tin(II) chloride is made by the action of dry HCl on tin, since chlorine would form the volatile $SnCl_4$. It readily takes up a molecule of water to form the pyramidal $SnCl_2.H_2O$. Tin(II) ions are extensively hydrolysed in aqueous solution, so tin(II) carbonate cannot exist. Excess acid must always be present to prevent hydrolysis with consequent precipitation of compounds like $Sn(OH)Cl.H_2O$. The action of sodium hydroxide on Sn(II) solutions precipitates hydrated tin(II) oxide, which redissolves in excess alkali forming stannites. A large range of halide complexes of Sn(II) are known, with formulae ranging from $SnCl^+$ to $SnCl_3^-$.

PbF_2, $Pb(NO_3)_2$ and $PbSO_4$ are rather similar to the equivalent salts of strontium, the sulphate being insoluble. Except for the strongly hydrolysed acetate, the other lead salts are insoluble and very often basic as well. Like tin, hydrated lead oxide is precipitated by the action of alkali on Pb^{2+} solutions: it is sufficiently acid to dissolve in a large excess of sodium hydroxide.

The complexity and interest of the chemistry of carbon lies in its remarkable capacity for forming bonds with itself, which leads to chains, rings and linked ring systems which very often contain hydrogen. These three types of structure can all incorporate other elements, principally N, O and S, thus providing an immensely intricate foundation for the emergence of a chemistry of living things. Some catenated compounds of silicon and germanium exist, but the chain length is strictly limited—unlike that of carbon, where compounds like polyethylene ($—CH_2—CH_2—)_n$ may have chain lengths of many thousands of carbon atoms. There are however no silicon analogues to the alkenes, alkynes and aromatic systems which help to give carbon chemistry its unique diversity.

The multiple bonds which carbon forms with other elements have no parallels in the later elements; silicon forms no analogues to the cyanide ion or carbon monoxide. CO_2 and SiO_2 are formally similar, but their properties could hardly be more different; the first a volatile gas soluble in water, the second an infusible solid which does not dissolve. Again, $R—C(OH)_2—R$ loses water to form $R—CO—R$, whereas $R—Si(OH)_2—R$ polymerises on dehydration to form silicones.

Certain carbon compounds will form ions whose existence is not limited to fractions of a second. The centre bond in $(C_6H_5)_3—C—C—(C_6H_5)_3$. will break to form the carbonium ion

$(C_6H_5)_3C^+$, the radical $(C_6H_5)_3C\cdot$ and the carbanion $(C_6H_5)_3C^-$, depending on the conditions. The $(C_5H_5)^-$ carbanion will form salts with electropositive metals. These compounds are not so much interesting for their own sake as for the light they cast on the existence of similar but less stable entities postulated as intermediates in all types of chemical reactions. Similar ions are formed much more reluctantly by the other M4 elements.

Like the remaining elements in M4, a single carbon atom can never occur as C^{4+}; but there is a potential C^{4-} ion in Al_4C_3, which contains discrete carbon atoms in the crystal lattice and hydrolyses to give methane. The reactions of CaC_2 are readily understandable in terms of a $^-C\equiv C^-$ ion, but its existence cannot be certain since calcium carbide is formed at a very high temperature and does not melt, nor does it dissolve in water without reaction. The carbides of the transitional elements are very stable and high melting; they play a vital role in metallurgy, particularly in the production of steel.

The halides of carbon cannot be made by direct action—indeed carbon itself is extremely inert. The bases of the charred poles used as uprights at Woodhenge, England, have remained unchanged for 4000 years. The series of compounds $C(Hal)_4$ becomes increasingly thermally unstable as the halogen changes from fluorine to iodine, yet all these compounds are highly resistant to hydrolysis and oxidation; $CHCl_3$ however, will oxidise in the presence of air and light to form $COCl_2$. The fluorocarbons have even higher thermal stability and chemical inertness than the analogous hydrocarbons. Those that are liquids are rather viscous, and the $(-CF_2-CF_2-)_n$ polymer has an exceedingly low coefficient of friction—hence the 'non-stick' frying pan.

A.5(5) GROUP M5

NITROGEN

PHOSPHORUS
ARSENIC
ANTIMONY
BISMUTH

Phosphorus, arsenic and antimony have an increasingly unstable yellow volatile allotrope with the formula X_4; the other allotropes are polymeric, and range from the non-conducting red phosphorus to metallic bismuth. The M5 elements are not attacked by dilute acids; phosphorus and arsenic yield hydrogen with concentrated alkalis, but antimony and bismuth do not react. All the elements except nitrogen burn in chlorine when finely divided. Down the Group the oxides become progressively more basic; $Bi(OH)_3$ is almost insoluble in alkali. The hydrides and alkyls lose their basic properties and become increasingly thermally unstable down the Group, and compounds displaying the Group oxidation state are strong oxidising agents. Nitrogen has little in common with the remaining elements in M5, and will be considered separately.

The binary compounds with M7 and M6 elements (excluding oxygen) are increasingly less volatile molecular crystals. The remainder range from bismuth alloys through infusible and unreactive solids to salt-like compounds with the highly electropositive elements which hydrolyse with dilute acids to generate the M5 hydrides. It is possible that Na_3P contains P^{3-}, but no other simple anions are known.

The hydrides are increasingly thermally unstable and increasingly strong reducing agents down the Group. PH_3 is a much weaker base than NH_3 and the other elements do not form XH_4^+ ions. The volatile alkyls are similar; phosphorus and arsenic have an extensive organic chemistry, but bismuth forms few compounds with carbon. Only phosphorus will form a compound of the type $H_2P—PH_2$, but all the elements will form $(CH_3)_2X—X(CH_3)_2$, bismuth with great reluctance.

All the trihalides exist and can be made (except PF_3) by direct combination with the M5 element in excess. They are increasingly less volatile and increasingly less hydrolysed by water: BiF_3 and $BiCl_3$ will conduct electricity when molten. All the pentafluorides exist, and except for PF_5 they are powerful fluorinating agents; none of the pentaiodides are known. The pentahalides of arsenic are notably less stable than those of phosphorus and antimony (p. 107).

The elements burn increasingly less readily in air to form the trioxides; phosphorus pentoxide is formed if the air is in excess. The other pentoxides can be prepared, with the possible exception of Bi_2O_5, by dissolving the elements in concentrated HNO_3 and evaporating gently.

Phosphorus(III) oxide is fairly soluble in cold water, forming the very weak phosphorous acid, but the other trioxides are increasingly insoluble; except for Bi_2O_3 they all dissolve in alkali to form oxyanions. Arsenic(III) oxide is just soluble in concentrated HCl, but the oxides of antimony(III) and bismuth(III) dissolve increasingly easily. These three acidic solutions will give highly coloured sulphide precipitates when H_2S is bubbled through them, which (except for Bi_2S_3) dissolve in solutions of alkali sulphides.

Phosphorus(v) oxide has the strongest avidity for water of any substance known; on boiling, orthophosphoric acid will be formed. Arsenic(v) oxide is freely soluble, antimony(v) oxide less soluble and bismuth(v) oxide completely insoluble (though its existence is not certain). The first three dissolve in solutions of alkalis to give salts; solid bismuthates are known. The solutions of arsenates and antimonates will precipitate the M_2S_5 compounds when H_2S is bubbled through them; the precipitate redissolves in excess alkaline sulphide to form thioarsenates and thioantimonates.

Orthophosphoric acid is a weak acid and contains tetrahedral PO_4^{3-} ions: under varied conditions a multitude of polymerised and condensed acids derived from this can be formed, but these are all hydrolysed back to orthophosphoric acid on boiling with moderately concentrated HNO_3. H_3AsO_4 forms few condensed acids, and $H(Sb(OH)_6)$ forms none at all. Phosphates are very difficult to reduce, arsenates are more easily reduced than antimonates $(Zn/H^+$ will convert AsO_4^{3-} to $AsH_3)$ while bismuthates are so easily reduced that they are classed as one of the strongest known oxidising agents, converting Mn^{2+} to MnO_4^- in strongly acid solution.

When As_2O_3 is dissolved in concentrated HCl, $AsCl_3$ is formed; but there is no evidence for the As^{3+} cation, and the salt is completely hydrolysed when the solution is diluted. Sb_2O_3 will dissolve in strong acid to give a hydrated sulphate which probably contains Sb^{3+}, but this is hydrolysed to $(SbO)_2SO_4$ on dilution. Bismuth(III) sulphate and nitrate are known, but the tendency to hydrolyse in solution to the insoluble oxysalts is still considerable.

PF_3 forms transitional metal complexes almost as readily as CO; R_3P and R_3PO also form strong bonds as ligands. PF_5 will form addition compounds with organic amines and ethers, and complex halides such as PF_6^-, PCl_6^-, AsF_4^-, $Sb_2F_7^-$, and BiI_4^- are all stable.

The hydride of nitrogen, and the NH_4^+ salts derived from it, are thermally more stable than the equivalent phosphorus compounds.

Nitride ions are more readily formed than phosphide ions and the nitride of titanium will conduct electricity when fused. These properties are to be expected from the trends in the remainder of M5.

However in almost every other way nitrogen and its compounds are entirely different from those of phosphorus. The element itself is gaseous, and much more inert than the phosphorus allotropes. There is no analogue for the $C\equiv N^-$ ion. The trihalides of nitrogen are all explosive liquids or gases which produce NH_3 on hydrolysis; no pentahalides are ever formed. The five oxides of nitrogen will all decompose on passing over hot copper, whereas the strongest reducing agents will scarcely affect the oxides of phosphorus. Nitrous acid has oxidising and reducing properties, phosphorous acid is only a reducing agent; nitric acid is a strong acid and strong oxidising agent, phosphoric acid is a weak acid with no oxidising powers. Salts containing the NO_3^- ion are all soluble and thermally unstable; salts containing the PO_4^{3-} ion are mostly insoluble and do not decompose on heating. NF_3 forms no complexes with the transitional metals whereas PF_3 forms a great number. The differences in the chemistries of these two elements could hardly be more glaring.

A.5(6) GROUP M6: OXYGEN AND SULPHUR

OXYGEN
_____

SULPHUR
Selenium
Tellurium
Polonium

Since the first member of the group is the commonest element in the earth's crust, and since the properties of the oxide of hydrogen dominate both biochemistry and geochemistry, the unique features of the chemistry of oxygen are of paramount importance. They differ extensively from those of sulphur, which has a tendency to catenate second only to that of carbon. Well-defined Group trends are shown by sulphur, selenium and tellurium; the latter elements are rare, but chemically interesting because their properties are regularly intermediate between those of their neighbours in M5 and M7.

The outstanding difference between the chemistry of oxygen and sulphur is the tendency of the latter to form chains. Oxygen has only one allotrope besides the O_2 molecule, highly reactive ozone. This and the corresponding O_3^- ion are the only compounds in which more than two oxygen atoms are bound to each other. Sulphur on the other hand readily forms chains and rings, and the interconversion of these leads to its complicated allotropy. Catenation is progressively less important in the chemistry of selenium and tellurium; Se_8 rings are formed, tellurium has a metallic appearance and polonium conducts electricity like a true metal.

The —O—O— bond is very weak. Hydrogen peroxide is an endothermic compound, organic compounds of the general formula R—O—O—H are dangerously explosive and ^-O_3S—O—O—SO_3^- and similar acid anions are very strong oxidising agents. In strong contrast, compounds H_2S_n (where n can be as high as 6) are common, the ^-S—S^- ion occurs regularly (e.g. in FeS_2, iron pyrites), thiols are easily oxidised to R—S—S—R, and ^-O_3S—S—S—SO_3^-, made by the action of iodine on thiosulphate ions, has no oxidising properties.

Oxygen will only show a positive oxidation state in its fluorides; at all other times its oxidation state is negative—apart from the peroxides it is always -2. Sulphur on the other hand shows oxidation states of -2 in the sulphides, $+2$(unusual) in SCl_2, $+4$ in SO_2 and SCl_4, $+6$ in SF_6 and SO_3. The latter combines vigorously with water, concentrated H_2SO_4 and HCl to form HO—SO_2—OH, HO—SO_2—O—SO_2—OH and HO—SO_2—Cl respectively.

The formation of hydrogen oxide from its elements is exothermic to the extent of 286 kJ mol^{-1} and it is only 1% dissociated at 2000°C; hydrogen sulphide can be formed in small yield at 400°C by passing hydrogen through molten sulphur (-21 kJ mol^{-1}) but a rise in temperature will cause it to decompose again. Water is a very weak acid and has no reducing properties, whereas H_2S is a slightly stronger acid and a strong reducing agent. The boiling point of water, like that of the hydrides of the neighbouring nitrogen and fluorine, does not fit into the Group trend.

The periodic properties of oxides have already been outlined (p. 23). Their similarities and differences from the sulphides are interesting. The compounds of both elements with electropositive metals are hydrolysed—as expected, since both are the salts of weak acids.

$$CaX + nH_2O \rightarrow (Ca^{2+} + X^{2-}) \rightarrow Ca^{2+}(aq) + XH^- + OH^-$$

The oxides and sulphides of the less electropositive metals frequently have different crystal structures and are non-stoichiometric; both sulphides and oxides are insoluble in water, but all the oxides and most of the sulphides dissolve in acids. The heat of formation of oxides is always higher than that of the equivalent sulphides, but for the later Main Group metals and metalloids the heat of formation of the sulphide is rather close to that of the oxide. The sulphides of these elements tend to be insoluble even in concentrated acids, which leads to their appearance in nature as sulphide ores—e.g. HgS, SnS_2, PbS, As_2S_3. Amphoteric character is common in oxides of these elements, and a few of their sulphides dissolve in alkali sulphide solution to form ions like AsS_3^{3-}. Sulphides of the non-metals tend to be volatile liquids, and are often more stable than the equivalent oxides.

The differences between oxygen and sulphur chemistry are so great that it is a relief to return to the more gradual Group trends exhibited by the elements sulphur, selenium, tellurium and polonium. Their hydrides are decreasingly thermally stable down the Group— more stable and more acidic than those of the equivalent M5 elements, but less stable and less acidic than the equivalent M7 elements. The dioxides are decreasingly volatile and decreasingly soluble and acidic; they are made by burning the elements in air, or dissolving them in nitric acid and evaporating. They are less stable and more acidic than the X_2O_3 oxides of M5, but more stable, less acid and less strong oxidising agents than the equivalent X_2O_5 oxides of M7. Strong oxidation of the lower oxides and oxysalts yields sulphates, selenates and tellurates, from which the free acids and the trioxides can readily be prepared. Selenic acid is very similar to sulphuric acid in its strength, dehydrating properties and formation of both normal and acid salts containing the SeO_4^{2-} ion; but it fits into the trend from AsO_4^{3-} to BrO_4^- in being thermally unstable and a very strong oxidising agent. Like arsenic (but unlike the lower Group members sulphur and phosphorus) there are practically no condensed or polymeric selenates. Normal and acid tellurates, derived from the weak dibasic acid $H_2(H_4TeO_6)$, are known, but they are only moderately strong oxidising agents.

The binary compounds of sulphur, selenium and tellurium with metals are all rather similar. They are often non-stoichiometric and decreasingly easily hydrolysed to the hydride by water and acids down any particular Group. The $X(Hal)_4$ compounds are increasingly thermally stable, and are decreasingly hydrolysed to the dioxides.

Selenium and tellurium will form $X(Hal)_6{}^{2-}$ compounds but sulphur will not. SF_6 is almost as stable as the noble gases, but SeF_6 and TeF_6 are increasingly reactive; the latter forms $TeF_7{}^-$, isoelectronic with IF_7.

There are traces of cationic behaviour for Te and Po. TeO_2, like Sb_2O_3, reacts with concentrated HCl to give $TeCl_4$, and some ill-defined basic sulphates have been reported. There appears to be no divalent ion intermediate between Sb^{3+} and I^+.

A.5(7) GROUP M7: THE HALOGENS

FLUORINE

CHLORINE
BROMINE
IODINE
Astatine

None of the halogens occur as elements in nature, for they are very reactive. The melting points and boiling points increase with atomic weight, and they exist as diatomic molecules in the vapour. They are prepared by the oxidation of naturally occurring halides. Fluorine behaves as a 'superhalogen' and will be considered separately.

Binary halides of the metals can always be made by direct combination, but indirect methods are sometimes necessary for preparing halides of non-metals. Iodine reacts less readily than the others, and only fluorine, chlorine and phosphorus react with it directly.

Chlorine tends to bring out higher oxidation states than bromine and iodine in the elements with which it combines. Thus PCl_5, $PbCl_4$ and ICl_3 all exist, whereas PI_5, $PbBr_4$ and IBr_3 do not. If the temperature is raised, halides of higher oxidation states will decompose, yielding the lower halide and free halogen.

The halides of the most electropositive metals are salt-like and high melting: they conduct electricity when molten, and they have a liquid range of many hundreds of degrees Centigrade. The boiling points of the $M2'$ halides and $Pb(Hal)_2$ compounds are below 1000°C, but there is still a long liquid range. Aluminium and the remaining Main Group metals have halides which boil at low temperatures very soon after melting, or actually sublime. The halides of the non-metals are mostly gases or volatile liquids at room temperature.

The halides of the most electropositive metals dissolve to give conducting solutions. The halides of the less electropositive metals react exothermically with water and are extensively hydrolysed in solution, while the halides of the non-metals react irreversibly with water, generating HHal.

With the electropositive metals the solubility of the halides generally falls from iodide to fluoride, but with certain elements at the end of the transitional Periods and in the later Main Groups the order is reversed (e.g. AgF is soluble whereas AgI is insoluble, and the same is true of the halides of lead). The same order holds for anionic halide complexes: with the electropositive elements the complex fluoride is the most stable (e.g. $AlF_6{}^{3-}$) but with the metals in M2′ and later Groups the complex iodide is the most stable (e.g. $HgI_4{}^{2-}$).

The hydrides can all be made by direct combination: the formation of HCl is explosive in sunlight, whereas the formation of HI is incomplete and requires a platinum catalyst. They can be made more conveniently by the hydrolysis of the P(III) halides or by the action of the non-volatile non-oxidising H_3PO_4 on metallic halides. HCl is unaffected by heat, whereas HI is 50% dissociated at 600°C; all the hydrogen halides react with metals at elevated temperatures. When passed into water, in which they are extremely soluble, much heat is evolved and the molecules are completely ionised according to the equation

$$HHal + H_2O \rightarrow H_3O^+ + Hal^-$$

In solvents like glacial acetic acid, HI ionises the most easily.

The hydrogen halides are increasingly strong reducing agents. HI will reduce concentrated H_2SO_4 to H_2S liberating iodine. Similarly the iodide ion is the most easily oxidised, even by such feeble oxidising agents as Cu^{2+}.

In their positive oxidation states the halogens are much more individualistic, and even the formulae of the various unstable oxides are not identical from one halogen to the next. Cl_2O_7 is the anhydride of perchloric acid and is the most stable of the chlorine oxides, but it detonates when heated or subjected to shock—as do the paramagnetic ClO_2 and ClO_3. The oxides of bromine are of very low thermal stability. Iodine forms the only solid oxides stable at room

P T—C

temperature. I_2O_5 is a strong oxidising agent and does not decompose until 300°C; I_2O_4 and I_4O_9 decompose more easily than this.

The halogens are increasingly insoluble in water, and increasingly reluctant to form hypohalites according to the equation

$$X_2 + 2H_2O \rightleftharpoons H_3O^+ + X^- + HOX$$

Adding alkali to remove the acid formed makes the forward reaction more favourable, but under these conditions the hypohalite will disproportionate to halate, IO^- more easily than ClO^-:

$$3XO^- \longrightarrow XO_3^- + 2X^-$$

Perhalates are made by electrolytic oxidation, but the stability of the bromine compound is very low. The perchlorates always contain the ClO_4^- ion; the periodates are of several types, IO_4^- and various acid salts of $H_3IO_6^{2-}$, all of which have strong oxidising power. The similar antimonates, tellurates, periodates and perxenates form a directly comparable series of compounds in their Group oxidation states. Their properties change regularly along the Period, e.g. their increasing power as oxidising agents.

The interhalogen compounds are low melting solids and liquids. They behave as oxidising agents, disproportionating and hydrolysing with comparative ease. Some polyhalide ions like ICl_4^- are derived from them; such ions generally exist only in the crystalline salts of large cations, but a few, like I_3^-, are stable in solution.

Iodine can show definite cationic properties. In the presence of concentrated HCl it can be converted by strong oxidising agents to ICl, which does not dissolve in CCl_4. The electrolysis of the dipyridyl iodonium ion $(C_5H_5N)_2I^+$ in anhydrous chloroform yields iodine at the cation only, and iodine is immediately liberated with iodide ions. Thus:

$$(C_5H_5N)_2I^+ + I^- \longrightarrow 2C_5H_5N + I_2$$

Much less stable bromine and chlorine analogues have been reported.

In acetic anhydride, fuming nitric acid will oxidise iodine to $I(OCOCH_3)_3$. When a solution of this compound in acetic anhydride is electrolysed, the amount of iodine liberated at the cathode corresponds to Faraday's Law calculations for I^{3+}. There are no analogous compounds of chlorine and bromine. The oxide I_4O_9 is also probably best formulated as $I^{3+}(IO_3^-)_3$.

Organic iodides are much more reactive than the analogous organic chlorides and bromides; they hydrolyse more easily and form

Grignard reagents more readily. Iodine will not add on to a simple $C=C$ bond, but it will displace hydrogen from a CH_3CO- group or a $-CO-CH_2-CO-$ group bonded to carbon.

Fluorine acts as a superhalogen: in some cases this merely leads to a continuation of the trends from iodine to chlorine, but in other cases a real discontinuity arises.

Fluorine is the most reactive of all the elements, and must be handled in an apparatus of copper or nickel, both of which form a protective layer of fluoride on the surface. It immediately decomposes water with the formation of HF.

Fluorine will always bring out the highest possible oxidation states of the elements with which it reacts: e.g. ICl_3 but IF_7, $FeCl_3$ but FeF_6. SF_6 is inert, but similar compounds in the later Periods, like UF_6, are violently corrosive gases.

Hydrogen fluoride is formed by the direct combination of the elements at $-259°C$ in the dark. It has a very high dielectric constant, which makes it an excellent solvent for ionic compounds. The anhydrous acid has a boiling point of $19°C$, a similar anomaly to the hydrides of nitrogen and oxygen. Anhydrous HF is an exceptionally strong acid: it will displace HCl from metallic chlorides, and will attack many stable oxides and oxyanions converting them to the equivalent fluoro compounds, e.g. with the silicates in glass it forms the SiF_6^{2-} ion. It is extremely corrosive to organic material and highly toxic. The pure acid scarcely conducts electricity, and when diluted it is only ionised to the extent of 1% according to the equation:

$$H_2O + 2HF \rightarrow H_3O^+ + FHF^-$$

The HF_2^- ion also exists in certain crystalline salts, KHF_2 for example.

Several unstable gaseous oxygen fluorides are known, but fluorine forms no oxyacids.

The organic chemistry of fluorine bears little relation to that of the other halogens. The C—F bond is exceptionally thermally stable and resistant to both chemical and biological attack: indeed the polymer $(-CF_2-CF_2-)_n$ is more stable than polyethylene. Fluorine itself will completely disrupt most organic compounds, so a metallic fluoride in a high oxidation state, CoF_3 for example, is used for fluorination.

A.5(8) GROUP M8: THE NOBLE GASES

Helium
Neon
Argon
Krypton
Xenon
Radon

The first chemical compound formed by the noble gases was prepared in 1962, when Xe and gaseous PtF_6 were allowed to react at room temperature and an orange solid $XePtF_6$ resulted. The oxidation state of xenon and the nature of the bonding in this compound are still obscure.

This discovery evoked an immediate response, and intensive work was started in laboratories in many parts of the world. The compound most easily made proved to be XeF_4, which condenses to a white solid when xenon and fluorine diluted with nitrogen are passed down a nickel tube at 400°C and the issuing gases are cooled. The compound is formed exothermically from its elements; it melts at 114°C and sublimes readily. Two other fluorides are known, which are also formed by direct combination. XeF_2 is a fairly stable crystalline solid melting at 140°C; XeF_6 is less thermally stable than the other fluorides and more reactive chemically, immediately attacking both glass and mercury. XeF_8 has been reported, but not confirmed.

Oxyfluorides and oxides are formed by the hydrolysis and disproportionation of the fluorides. XeO_3 occurs as white, non-volatile, highly explosive crystals, while XeO_4 is a yellow solid unstable at room temperature. The aqueous solution chemistry of xenon is extensive and somewhat complicated. The aqueous hydrolysis of XeF_4 leads to xenon and compounds of Xe(VI) and the latter can be precipitated as Ba_3XeO_6 with barium carbonate: it is a strong oxidising agent. Xe(VIII) cannot be prepared in acid solution but is transitorily stable in alkalis; it is usually prepared by the hydrolysis of XeF_6 with sodium hydroxide which is followed by disproportionation. Various salts are known, Na_4XeO_6 being typical; they are all exceedingly strong oxidising agents.

The compounds of krypton are less stable than those of xenon. KrF_2 is synthesised by the irradiation of the mixed gases with ultraviolet light when suddenly chilled to $-253°C$, and KrF_4 has been formed by passing stoichiometric amounts of the elements through an electric discharge between copper electrodes. The latter compound decomposes quite rapidly at room temperature. It hydrolyses and disproportionates in solution to give a derivative of $Kr(VI)$ which can be precipitated as $BaKrO_4$.

The compounds of radon are very difficult to isolate owing to the short half-life of this element and the special techniques required for dealing with very small quantities of highly radioactive substances. A fluoride has been prepared but its formula is uncertain.

A.5(9) TRANSITIONAL ELEMENTS OF THE FIRST LONG PERIOD

(a) All the transitional elements of the first Period are hard high melting point metals, increasingly dense and decreasingly electropositive across the Period. The lower oxides are infusible and often coloured; they dissolve readily in acids to give hydrated cations and the salts are extensively hydrolysed in aqueous solution.
(b) Isomorphous compounds are very common: e.g. $XSO_4.7H_2O$ where X may be Cr, Mn or Fe.
(c) All transitional elements exist in several oxidation states. The oxidation state of $+2$ becomes increasingly stable across the Period with respect to higher oxidation states: for example Ti^{2+} will reduce water, while Cu^{2+} will oxidise I^- to I_2. The maximum value of the highest oxidation state falls uniformly from manganese onwards.
(d) A large number of both cations and anions containing transitional metals are strongly coloured in aqueous solution: for example Fe^{2+} is green, Fe^{3+} is brown; MnO_4^- is purple, MnO_4^{2-} is dark green.
(e) Solutions of transitional metal ions often act as homogeneous catalysts while transitional metals and their insoluble oxides are effective heterogeneous catalysts. Co^{2+} will catalyse the oxidation of a solution of H_2O_2 and so will solid MnO_2; finely divided nickel will catalyse the hydrogenation of oils to fats.
(f) At least some of the ions formed by all the transitional metals are paramagnetic.

(g) All transitional elements will form a multitude of cationic, neutral and anionic complexes with other stable molecules or ions. Thus:

$$Cu^{2+} \text{ will form } Cu(NH_3)_4{}^{2+} \text{ with ammonia}$$
$$Ni \text{ will form } Ni(CO)_4 \text{ with carbon monoxide}$$
$$Fe^{2+} \text{ will form } Fe(CN)_6{}^{4-} \text{ with cyanide ions.}$$

A.6
Special topics

A.6(1) DIAGONAL RELATIONSHIPS

There are close resemblances between elements related diagonally to one another in the early Main Groups of the Periodic Table. The most striking similarities are as follows:

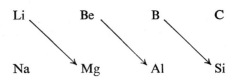

a   Lithium/Magnesium

(a) Instability of both carbonates and hydroxides to heat.
(b) Sparingly soluble fluorides, carbonates and phosphates.
(c) Both elements combine readily with molecular nitrogen at 300°C.
(d) The salts of both elements are hydrated.
(e) Both readily form volatile alkyls.

But
(a) The crystal structures of salts of the same anions are necessarily different.
(b) Lithium readily forms a salt-like hydride, whereas the hydride of magnesium is much less stable and does not show salt-like properties.
(c) Lithium hydroxide (solubility 4×10^{-3} mol. cm^{-3}) is a little less soluble than sodium hydroxide, but vastly more soluble than magnesium hydroxide (solubility 7×10^{-7} mol. cm^{-3}).

b   Beryllium/Aluminium

(a) The elements have similar reactions to acids and alkalis.
(b) Both the oxides and hydroxides are amphoteric, and stable beryllates and aluminates are formed.
(c) Both chlorides are rather volatile, and are excellent Friedel–Crafts catalysts.
(d) Both chlorides are extensively hydrolysed in solution.
(e) The hydrides and methyls are both high melting and presumably polymeric.

But

(a) Hydrated beryllium salts contain only $4H_2O$, whereas most hydrated aluminium salts contain $6H_2O$.

(b) There is no BeH_4^{2-} corresponding to AlH_4^-.

(c) Beryllium compounds are extremely toxic, whereas aluminium and its compounds are generally harmless to life.

c Boron/Silicon

(a) High melting point non-metals, inert to dilute acid and alkali but dissolved slowly by hot concentrated alkali.

(b) Both elements form many condensed and polymerised oxyanions.

(c) The hydrides are volatile, pyrophoric and completely hydrolysed in water.

(d) The chlorides are both volatile and easily hydrolysed.

(e) The fluorides are partially hydrolysed, but readily combine with the fluoride ion so formed.

But

(a) Boric acid, $B(OH)_3$, is moderately soluble in water, whereas hydrated silicon dioxide is not appreciably soluble.

(b) BF_3 forms numerous adducts with alcohols, ether and amines, whereas SiF_4 does not.

(c) The fluoride complex anions of the two elements have different structures: BF_4^-, but SiF_6^{2-}.

(d) There are no silicohydride ions to correspond to the borohydrides BH_4^-.

A.6(2) CONTRAST OF THE FIRST TRANSITIONAL SERIES WITH THE SECOND AND THIRD TRANSITIONAL SERIES

The elements of the first transitional series differ sharply in chemical properties from the elements belonging to the same Groups in the second long Period, whereas the second and third series transitional elements are rather similar [see A.6(3)].

The contrast is most conveniently summed up under the following headings:

a Decreasing electropositivity of metals

The early elements in the first transitional series from titanium to iron liberate hydrogen with dilute acids. Elements of the second transitional series are much less electropositive and do so less readily.

b   Decreasing importance of cations

Simple cations which form stable salts are much more frequently found in the first transitional period. For example Mn^{2+} is stable whereas Tc^{2+} and Re^{2+} are almost unknown. Again, Ti^{2+} and Ti^{3+} have no analogues in the chemistry of Zr and Hf.

In the higher oxidation states compounds like VO^{2+} are sometimes formed; in this way too high a charge on the cation is avoided.

c   Increasing importance of anionic complexes

Titanium does form some titanates and complex halides, e.g. $TiCl_6^{2-}$, but Zr and Hf are never found in any other form. Platinum forms the very stable $PtCl_6^{2-}$ ion, for which there is no nickel analogue.

d   The increasing stability of higher oxidation states

There are many cases of this trend; here are three notable examples. $+3$ is the most stable oxidation state of chromium but $+6$ is the most stable oxidation state of molybdenum and tungsten; thus chromates are strong oxidising agents, whereas molybdates and tungstates are not. The $+3$ state of gold is much more stable than the $+3$ state of silver and copper. There is no FeO_4, but RuO_4 and OsO_4 are both known.

A.6(3) SIMILARITY OF THE SECOND AND THIRD TRANSITIONAL SERIES

The properties of the second and third transitional series are very similar, and rather different to those of the first [see A.6(2)].

This similarity if most marked in the case of T4 and T5. In the former, the properties of Zr and Hf are almost identical, except for tiny differences in solubility and volatility, and the same is largely true for Nb and Ta. Thus it is natural that these elements should occur together in nature, and it is only since the advent of adsorption chromatography that efficient separations have been achieved.

The elements of the third Period are even less electropositive than those of the second, but nevertheless the stabilities of the comparable oxidation states are very similar. Both ruthenium and osmium form a tetroxide and unlike nickel both palladium and platinum show a stable oxidation state of $+4$ as well as $+2$. There is a tendency towards the end of the third Period to form oxidation states higher than those in the second Period—witness the comparatively stable $AuCl_4^-$ ion—and PtF_6, although very reactive, has no palladium analogue.

A.6(4) THE CHEMISTRY OF THE INNER TRANSITIONAL ELEMENTS: THE LANTHANIDES AND ACTINIDES

The lanthanides from cerium to ytterbium are prepared by the electrolysis of their fused chlorides because they are highly electropositive metals, but the electropositivity decreases across the Period. An alloy of the mixed metals can be used as a lighter flint, because their tendency to combine with oxygen is very great.

For these elements the oxidation state of $+3$ is far more important than any other. Their chemical properties and the solubilities of their salts are very similar indeed, and they are best separated by chromatographic methods. Cerium shows an oxidation state of $+4$ in the orange cerium(IV) sulphate which is used as a volumetric oxidising agent. Some of the other elements form compounds in oxidation states of $+2$ and $+4$, but these are all unstable with respect to the $+3$ state, especially towards the end of the Period.

All the actinides, from element 90, Th, to element 103, Lw, are radioactive, and all of them after uranium are made artificially. Because of rarity, their chemistry is studied by carrier techniques and reactions in interconnected capillary tubes.

Like the lanthanides, the actinides are very electropositive. Early in the actinide series, however, there is a very much greater variety in the oxidation states shown, and like the later transitional series higher oxidation states are more favourable; the $+3$ state becomes increasingly important across the Period, and curium has only a limited tendency to form the $+4$ state. Uranium on the other hand has stable oxidation states of $+3$, $+4$ and $+6$ in the ion UO_2^{2+} and in UF_6. The latter is a violently corrosive gas and is of interest because it was used to separate U^{235} from U^{238} for the first atomic bomb by the method of gaseous diffusion. Actinides tend to form more complexes than lanthanides, and their salts are hydrolysed to a greater extent. The solubilities of salts in the $+3$ oxidation state are very similar to those of the lanthanide in the same Group.

A.7
Experimental demonstration of group properties

A.7(1) GROUP M1: THE ALKALI METALS

Lithium is too expensive for general laboratory use.

1 Action of water on the elements (CARE)

Dry a small chunk of sodium in a filter paper, and wrap in a thin copper gauze weighted with lead. Drop this package into a large beaker of water. Repeat with a similar sized lump of potassium. Is it clear which of the two metals is the most reactive?

If a piece of lithium is available, it is very striking to see how relatively slowly it reacts with water compared to the other M1 elements.

2 Thermal stability of oxysalts

Carefully melt a dry sample of potassium nitrate in an ignition-tube, then raise the temperature and test for the gas evolved. Repeat for potassium perchlorate. Why are potassium salts more often used as laboratory reagents than sodium salts?

Attempt to melt samples of potassium carbonate and sodium orthophosphate. Comment on their ease of melting and any gases which may be evolved.

Acid salts are less thermally stable than normal salts

Dissolve a little solid sodium hydrogen carbonate in half a test-tube of water by shaking, and add a few drops of universal indicator. Pour the solution into a small beaker and boil for a few minutes. Cool, and add a few more drops of indicator. Can you account for the results?

4 Lattice energy of salts

To a saturated solution of sodium perchlorate add dilute potassium chloride solution. Can you explain what happens? Repeat the experiment using dilute ammonium chloride solution; does this confirm your original hypothesis?

5 Reactions of alkali peroxides

Burning sodium in air is rather hazardous, so use commercial sodium peroxide. Dissolve it in water, and test the pH of the resulting solution with universal indicator. What will be formed if the solid is added to cold dilute HCl solution? Is there any way in which you can demonstrate this?

6 Flame spectra for distinguishing the M1 Metals

Take a small portion of a sodium salt, dampen it with concentrated HCl solution (chlorides are more volatile than other salts) and dip a nickel wire into the slurry and hold it in the colourless outer flame of a Bunsen burner. Repeat the experiment with a pure potassium salt and a clean nickel wire free from contamination with sodium salts, and view the flame colour through blue glass. What is the function of the blue glass in this experiment?

7 Comparison of potassium and copper

a Reactions with water

Drop a *very* small piece of the metals into water and observe whether any gas is evolved; test the resulting liquid with phenolphthalein.

b Thermal stability of sulphates

Heat the sulphates strongly in ignition-tubes and test for the gases evolved.

c Reactions of the ions in solution

To separate solutions of the sulphates add (a) litmus solution, (b) sodium carbonate solution, (c) sodium hydroxide solution, (d) ammonium hydroxide solution, (e) hydrogen sulphide solution, (f) sodium perchlorate solution. Comment upon any differences observed.

A.7(2a) GROUP M2: THE ALKALINE EARTHS

Beryllium is too toxic for general laboratory use.

1 Action of water on the elements

Compare the action of half a test-tube of warm water on (a) clean metallic calcium, (b) a freshly sandpapered strip of magnesium ribbon.

2   Thermal stability of nitrates

Carefully drive off the water of crystallisation from separate samples
of barium and magnesium nitrate, then heat both salts strongly in an
ignition-tube. Which of the two nitrates most resembles sodium
nitrate in its decomposition? Can you interpret your results in terms
of electropositivity?

Repeat the experiment with barium and magnesium carbonates;
which of these two salts decomposes most easily?

3   Lattice energy of salts

To separate test-tubes containing about 3 cm³ of dilute solutions of
barium chloride and magnesium chloride, add (a) sodium fluoride
solution, (b) dilute sulphuric acid, (c) lime water, (d) sodium
carbonate solution. Interpret any differences between the reactions
of the barium and magnesium compounds.

4   Salts of weak acids

Prepare three test-tubes containing respectively dilute barium chloride
solution with its own volume of (a) water, (b) dilute acetic acid, (c)
dilute hydrochloric acid. Divide the contents of each test-tube into
two parts; to the first part add potassium chromate solution, to the
second, add ammonium oxalate solution. Are the results significant?

5   Flame spectra for distinguishing the M2 metals

Examine the flame spectra for salts of the M2 metals (p. 56). Can you
explain why magnesium salts do not apparently impart colour to the
flame?

6   Complex formation

Compare the solubilities of anhydrous magnesium perchlorate and
anhydrous barium perchlorate in (a) water, (b) ethanol, (c) diethyl
ether.

Repeat experiment A.7(2a)3d using a mixture of dilute solutions of
ammonium chloride, ammonium hydroxide and ammonium
carbonate. Do you notice any difference?

7   Comparison of calcium and zinc

a   Action of water on the elements
Cover samples of the clean granulated metals with 3 cm³ of water.
Test for the gas evolved, and add a few drops of phenolphthalein
to the solutions.

b   Action of heat on the elements

Put a few grains of the metals on the lids of porcelain crucibles and attempt to ignite them from above with the tip of a Bunsen flame.

c   Reaction of the oxides with water

To about one gram of the freshly prepared oxides add 3 cm³ of water. Which oxide liberates the more heat and which of the resulting solutions is the more alkaline?

d   Action of halogens

Add a drop of bromine to a small piece of the granulated metals in two test-tubes. Is there any sign of reaction? Add a little water, filter and add sodium carbonate solution to the filtrate. What is the precipitate which is formed?

e   Action of the metals with sulphur

To a small portion of the granulated metals in ignition-tubes add a little sulphur and warm. After the violent reaction has finished, sublime off the excess sulphur, allow the tube to cool and add a few drops of dilute acid. What has happened?

f   Reaction of the ions in solution

To 3 cm³ of solutions of the colourless nitrates add (a) sodium carbonate solution, (b) sodium hydroxide solution drop by drop, (c) sodium sulphate solution, (d) hydrogen sulphide solution, (e) potassium chromate solution, (f) potassium hexacyanoferrate(II) solution. Comment on any differences you observe.

A.7(2b) METALS IN THE LATER MAIN GROUPS

1   Sulphides which are mostly insoluble in acids

Prepare solutions of the following, add a little dilute hydrochloric acid, then dampen a filter paper with them and expose the damp patch to a stream of H_2S from a Kipp's Apparatus: (a) $SnCl_2$, (b) $BiCl_3$, (c) $Pb(NO_3)_2$, (d) $ZnCl_2$, (e) $Cd(NO_3)_2$, (f) $Hg(NO_3)_2$, (g) $SbCl_3$. Record the colour of the precipitate, if any, and explain your results.

Attempt to dissolve these precipitates using yellow ammonium sulphide solution. Is there any pattern in your results?

2   Stability of anionic complexes

Precipitate the iodides of mercury(II) and lead(II) by adding a few drops of dilute potassium iodide solution to 3 cm³ of solutions of the nitrates. Allow the precipitates to settle and pour off the supernatant liquid, then add an excess of saturated potassium iodide solution. To the resulting liquids add a dilute solution of sodium fluoride.

Compare your result with that obtained by mixing magnesium chloride solution with potassium iodide solution, then adding sodium fluoride solution.

3   Insolubility of halides

Add a few drops of dilute hydrochloric acid to a solution of lead(II) nitrate, then add 3 cm³ of concentrated hydrochloric acid, warm and filter. On the resulting filtrate try the action of (a) a few drops of potassium iodide solution, (b) a solution of hydrogen sulphide.

Slowly bring to the boil a saturated solution of lead(II) iodide containing a small amount of excess solid. Cool the solution so formed under a cold tap.

4   Decomposition of sulphur-containing complexes

Slowly add sodium thiosulphate solution drop by drop to a solution of lead(II) nitrate in a beaker until there is a bulky precipitate. Boil the beaker vigorously until the precipitate turns black, then filter the suspension and add barium nitrate in dilute nitric acid to the filtrate. Interpret your results and write an equation for the decomposition.

A.7(2c) GROUP M2′: ZINC, CADMIUM AND MERCURY

1   Electropositivity of metals

Try and dissolve a piece of granulated zinc first in cold then in hot dilute sulphuric acid.

Add 2 drops of dilute copper(II) sulphate solution to the reaction mixture, and observe what happens.

2   Hydrolysis of salts

To solutions of the nitrates of zinc, cadmium and mercury(II), add universal indicator solution. Then add sodium hydroxide solution dropwise until it is present in excess. Are any of the hydroxides amphoteric? See if the result is similar if ammonium hydroxide is used instead of sodium hydroxide, and explain any differences.

3 Solubility of fluorides

Magnesium fluoride is precipitated by adding a solution of sodium fluoride to a solution of a magnesium salt. What is the action of sodium fluoride solution on solutions of the chlorides of M2'.

4 Relative stability of complexes

(a) Pass some H_2S from a Kipp's Apparatus into a solution of zinc hydroxide redissolved in excess alkali. Is there a white precipitate of ZnS?

(b) Redissolve in excess the scarlet precipitate HgI_2 formed by adding potassium iodide solution to mercury(II) nitrate solution. Dampen a filter paper with the clear liquid, then hold it above a strong ammonia solution.

(c) Compare the strength of the complexes formed by unsaturated ligands with transitional metals and M2' metals. Precipitate CuS and CdS from separate solutions of their salts. To two fresh salt solutions add potassium cyanide solution (CARE) until the precipitates at first formed disappear; then pass H_2S through the resulting solutions and interpret your results.

(d) Form the $Hg(CNS)_4^{2-}$ complex ion by adding excess ammonium thiocyanate solution to mercury(II) nitrate solution. Try and precipitate HgI_2 by adding potassium iodide solution, and HgS by passing H_2S gas. Is the mercury(II) thiocyanate complex stable?

5 Reactions of the Hg_2^{2+} ion

(a) Add a few drops of potassium iodide solution to a solution of mercury(I) nitrate, watching very carefully to see if any transient colour appears. To interpret what has happened, is it important to know that HgI_2 is very insoluble? Now add excess saturated potassium iodide solution to the mixture and filter off the precipitate. Can you say what it is?

(b) Add a few drops of mercury(I) nitrate solution to excess tin(II) chloride solution and observe what happens. What is the grey precipitate that is finally formed?

(c) To a solution of mercury(I) nitrate, add first a few drops of ammonium thiocyanate, then a large excess. Comment on your results.

A.7(3) GROUP M3: BORON AND ALUMINIUM

1 Borate esters

Dissolve some borax, $Na_2B_4O_7$, in ethanol in an evaporating dish, then add concentrated sulphuric acid and warm. $B(OC_2H_5)_3$ is formed and burns with a characteristic green flame.

2 Electropositive nature of aluminium

(a) Try to dissolve aluminium foil in water, then in dilute sulphuric acid, then in dilute sulphuric acid containing three drops of copper(II) sulphate solution.
(b) Take a similar piece of aluminium foil, dip it momentarily into mercury(II) chloride solution, immerse it in distilled water. Account for the result.
(c) Intimately mix some aluminium powder and some finely ground sulphur. Fill an ignition-tube to a depth of one inch with the mixture, clamp it securely in a fume cupboard pointing in a safe direction, then heat it from below, beginning at the top of the solid. The reaction is extremely exothermic! (CARE)

3 Amphoteric nature of aluminium

(a) Devarda's Alloy is mostly aluminium, activated by alloying it with copper. Try the reaction of concentrated sodium hydroxide solution on the solid, and test for the gas evolved. Filter the mixture into a solution of half dilute/half concentrated sulphuric acid, making sure the acid remains in excess throughout; put away to cool and observe the crystals formed.
(b) To a solution of aluminium potassium sulphate made from the crystals, add dilute sodium hydroxide solution drop by drop until the solution is alkaline. Comment upon what happens.

4 Covalent compounds of aluminium

To a small amount of finely powdered aluminium in a test-tube clamped in a fume cupboard add two drops of liquid bromine from a dropping tube and note the great heat evolved. Now attach an open delivery tube and heat the solid. A fine powder will sublime from the excess unchanged metal and deposit in the cool part of the delivery tube. Test the solubility of this compound in diethyl ether; then to another portion add a few drops of water and compare the results.

5 Hydrolysis of aluminium salts

(a) Test a solution of aluminium sulphate with universal indicator.
(b) Attempt to prepare aluminium carbonate by adding a solution
of dilute sodium carbonate solution to aluminium sulphate solution.
What gas is given off? Can you deduce what the precipitate must be?

A.7(4) GROUP M4

1 Hydrolysis of calcium carbide

(a) Cover a few lumps of calcium carbide in a beaker with 3 cm³ of
water. What is the gas that is formed, and what impurity makes it
spontaneously inflammable? Test the solution with phenolphthalein,
and write an equation to account for your results.
(b) Fill a gas jar with chlorine, then introduce 3 cm³ of water followed
by a few small lumps of calcium carbide. Account for what you observe.

2 Formation and hydrolysis of magnesium silicide

Cut about ten strips of magnesium ribbon 2 cm in length and
introduce them one by one into a glass test-tube held horizontally
in a roaring Bunsen flame (glass contains a high proportion of SiO_2).
The magnesium melts and becomes stuck to the glass which is slowly
rotated until all the magnesium ribbon has been introduced and the
bottom inch of the test-tube is almost entirely coated with it. Cork
the tube and allow it to cool, then introduce 3 cm³ of water; a gas
spontaneously inflammable in air is evolved.

3 Action of hydrogen fluoride on silica

Pare a few shavings from a wax candle and pile them on a watchglass, then allow the watchglass to stand in an oven until the wax
melts and thinly coats the surface of the glass. Pour off any excess
wax and allow to cool. Write an inscription with a stylus, making
sure that the wax is pierced, then fill the glass with a concentrated
solution of hydrogen fluoride from a resistant plastic bottle (CARE)
and leave in a fume cupboard for a few days. Pour off the excess
acid, then melt the wax in the oven and pour it off from the glass.
The inscription, etched into the glass, can be clearly seen.

4 Formation of silica gel from waterglass (sodium silicate)

Dissolve some waterglass in several times its own volume of hot
water and, when cool, titrate with M hydrochloric acid using
phenolphthalein as indicator. Allow the final product to set into a gel.

5 Amphoteric nature of tin(IV) sulphide

Dissolve some hydrated tin(IV) oxide in concentrated hydrochloric acid, filter the solution and pass in a stream of H_2S from a Kipp's Apparatus. Filter off the yellow precipitate of SnS_2, wash the filter paper with dilute hydrochloric acid, then pour some yellow ammonium sulphide solution over the precipitate and preserve the filtrate. What is the colour of the filtrate and what happens to it when it is acidified?

6 Formation of tin(IV) bromide: Reaction as a Lewis Acid (p. 104)

(a) To a few lumps of granulated tin in a test-tube add 5 drops of bromine, making sure that the tin is always in excess. Distil off the volatile solid that is formed into a clean test-tube. Dissolve the solid in carbon tetrachloride.
(b) To this solution add a very little crystal violet indicator and note the colour. What is the effect of passing dry ammonia gas through one portion of the solution, and adding diethyl ether to another portion?

7 Reducing power of Sn^{2+}

To a solution of mercury(II) chloride in dilute hydrochloric acid add an excess of tin(II) chloride solution. At first a white precipitate of mercury(I) chloride is formed, but this gradually turns grey. How do you account for the change?

8 The halides of lead

(a) Filter some cold saturated lead(II) chloride solution and add potassium iodide solution to the filtrate. Which is more soluble, lead(II) chloride or lead(II) iodide?
(b) Boil a sample of lead(II) chloride with water in a test-tube and filter it into another test-tube. Cool under the tap and observe what happens. Is lead(II) iodide more soluble in hot than cold water also?
(c) To the cold suspension of lead(II) chloride from the last experiment, add 2 cm³ of concentrated hydrochloric acid. Does the precipitate disappear? Warm the solution and cool it again; are there any further changes?
(d) Can you predict whether a suspension of lead(II) iodide will dissolve in concentrated hydrochloric acid? Check your prediction.

9 Oxidation of lead(II) sulphide

Damp a filter paper with lead(II) nitrate solution, and expose it to H_2S gas over an uncorked bottle of the solution. On the brown stain of lead(II) sulphide so formed, try the effect of adding 20 volume H_2O_2 from a dropping tube. What is formed in this reaction?

10   Oxidation of an alkaline solution of Pb^{2+}

To a solution of lead(II) nitrate add a solution of sodium hypochlorite, (made by dissolving chlorine in sodium hydroxide solution). Filter off the brown precipitate formed. What is it? What gas is liberated when it is dissolved in concentrated hydrochloric acid?

A.7(5)   GROUP M5

1   Formation of the chlorides of the elements

(a) Fill a burette with dry chlorine by upward displacement. Cork it, and in a fume cupboard open the tap under a watchglass filled with concentrated ammonia. Cool the burette by stroking it with a piece of cotton wool damped with ether so that the gas it contains contracts, and about 1 cm³ of liquid is sucked up into the base of the burette. Close the tap, withdraw the burette and invert it rapidly. Nitrogen(III) chloride is formed, but immediately decomposes with a flash of light.

(b) Introduce a small dry lump of white phosphorus in a deflagrating spoon into a gas jar of dry chlorine. The phosphorus will immediately inflame, and a white solid (PCl_5) will be deposited on the sides of the gas jar.

(c) Sprinkle grey powdered antimony into a gas jar of chlorine. The particles inflame, and burn to a white powder, which hydrolyses in damp air if the gas jar is left open.

(d) To warm moderately concentrated hydrochloric acid add bismuth oxide until there is a slight permanent precipitate. Filter the solution into a boiling-tube of distilled water. What is the permanent precipitate that forms? Will it dissolve in concentrated hydrochloric acid?

2   Formation of the hydrides of the elements

(a) Put approximately 1 g of magnesium filings in a crucible with a lid, support it on a pipeclay triangle and bring it to red heat, carefully removing the lid at intervals and allowing as little as possible of the smoke to escape. When all the magnesium metal has reacted allow the crucible to cool, then dissolve the white powder which has been formed in the minimum amount of moderately concentrated hydrochloric acid. Evaporate until only a small quantity of liquid is left, cool, and add concentrated sodium hydroxide with a dropping tube. Test the gases that escape first with litmus paper and then with Nessler's reagent. It is important that no splashes should contaminate the test papers.

(b) Mix a small quantity of solid sodium orthophosphate with six times its weight of finely powdered magnesium. Put the mixture into a crucible with a lid and heat gingerly, and then strongly. Allow to cool, then add water and test the gases evolved with universal indicator paper, and a strip of filter paper dipped in concentrated copper(II) sulphate solution.

(c) Fit a test-tube with a two-hole cork, through which pass a teat-pipette containing a few drops of sodium arsenate solution and a hard glass delivery tube terminating in a jet. In the test-tube put some arsenic-free zinc and some dilute sulphuric acid and warm; when the solution is emitting hydrogen freely (a crystal of copper(II) sulphate may be required) burn the gas issuing from the jet in a Bunsen flame. Put another lighted Bunsen under the delivery tube half way along, then pump a few drops of the arsenic-containing solution into the reacting mixture. AsH_3 is formed, and decomposes in the hot zone of the delivery tube; arsenic is deposited on the cooler walls of the tube towards the jet in a characteristic metallic stain.

3 Action of nitric acid on the elements

In two boiling-tubes fitted with reflux condensers place about 1 g of red phosphorus and 1 g of bismuth; cover both with 3 cm³ of concentrated nitric acid. Reflux for half an hour, then filter; collect the filtrate in distilled water. Can you interpret the differing behaviour of these two elements?

4 Stepwise hydrolysis of phosphorus pentoxide

(a) Dissolve 1 g of phosphorus(V) oxide in about 25 cm³ of distilled water in a conical flask. After swilling for a few minutes pour off a small portion and add dilute silver nitrate solution. Look up other tests which confirm the presence of a high molecular weight phosphate polymer.

(b) Simmer half the remaining solution for twenty minutes, being careful not to allow it to boil vigorously; cool and add dilute silver nitrate solution. Is there any change in the precipitate formed?

(c) Reflux the remaining solution vigorously for an hour. Cool and add dilute silver(I) nitrate solution. How do you explain the result?

5 Test for the —OH group using phosphorus pentachloride

To boiling-tubes in a fume cupboard containing 3 cm^3 of water, ethanol, acetic acid, acetone and diethyl ether add a very small sample of solid phosphorus(v) chloride. Some of the liquids react violently and produce extensive fumes; is this significant?

6 Sulphides of the elements

Through solutions of arsenic(III) chloride (CARE), potassium antimonyl tartrate and bismuth(III) chloride, pass H_2S gas from a Kipp's Apparatus. Filter off the coloured precipitates so formed, wash them, and attempt to dissolve them in yellow ammonium sulphide solution. Are the results so obtained in accordance with the metallic nature of the elements?

7 Oxidising power of sodium bismuthate

Mix a few drops of manganese(II) sulphate solution with 5 cm^3 of concentrated nitric acid diluted with 5 cm^3 of water. Sprinkle very sparingly a few grains of sodium bismuthate into the solution and watch for the appearance of a purple stain in the wake of the sedimenting particles. What is happening?

A.7(6) GROUP M6: OXYGEN AND SULPHUR

1 Comparison of the reactivity of H_2O and H_2S with metals

Partially amalgamate the surface of two strips of zinc foil by rubbing with some dilute mercury(II) chloride solution. Compare the action of steam and H_2S gas when passed across the foil in a heated combustion tube. Perform a similar experiment with heated copper foil. What can you deduce?

2 Comparison of the thermal stability of mercury(II) oxide and mercury(II) sulphide

Strongly heat very small samples of mercury(II) oxide and mercury(II) sulphide in ignition-tubes. Can you say what the products are? Are the results similar?

3 Action of oxides and sulphides with acids

(a) Compare the action of water, dilute acetic acid and dilute hydrochloric acid on samples of zinc oxide and zinc sulphide.
(b) Test a solution of H_2S water with universal indicator paper then

add 3 cm³ of dilute copper(II) sulphate solution and test again. Can you account for what has happened?

4 Comparison of solutions of ammonium hydroxide and ammonium sulphide

Compare the action of ammonium hydroxide and ammonium sulphide solutions when added drop by drop to test-tubes of (a) zinc sulphate solution, (b) potassium antimonyl tartrate solution. Note the similarities and differences.

5 Dehydrating power of thionyl chloride (no oxygen analogue)

Add 2 cm³ of $SOCl_2$ to a small sample of hydrated cobalt(II) chloride in a boiling-tube clamped in a fume cupboard. Periodically take the temperature of the mixture and observe the colour change. Gently warm the boiling-tube to drive off excess thionyl chloride; what is the solid which remains?

This is one of the very few spontaneous endothermic reactions which is easily demonstrated.

A.7(7) GROUP M7: THE HALOGENS

Fluorine itself is too difficult to prepare and too toxic for general laboratory use. Anhydrous HF gives rise to very serious burns if it comes in contact with the skin.

1 Direct combination of halogens with the elements (CARE)

a) Phosphorus: see A.7(5) 1(b).
(b) Antimony: see A.7(5) 1(c).
(c) Tin: see M4, Expt. 6.
(d) Mercury. Introduce a very small droplet of mercury into a clean dry ignition-tube and add several lumps of iodine, then heat the mixed solids. Excess iodine will sublime as a violet vapour, leaving scarlet HgI_2 behind.

2 Solubility of halogens

(a) Pour a little liquid bromine into 100 cm³ of water in a flat bottomed flash, cork and leave for 30 minutes. Can you tell if any bromine has dissolved? Now add 50 cm³ of CCl_4 and shake: which layer contains the majority of the bromine? From these results devise a test for distinguishing the brown vapours of bromine and nitrogen dioxide.

(b) Shake a lump of iodine with distilled water in a corked test-tube; is the iodine soluble? Now add some concentrated potassium iodide solution and shake again. Can you explain what has happened?
(c) Divide the solution from Expt. 2(b) into two halves. Shake the first portion with an equal volume of diethyl ether, the second with carbon tetrachloride. Is the iodine more soluble in the organic layers? Can you explain the colour changes involved?

3 Displacement of one halogen by another

(a) Add about 1 cm^3 of CCl_4 to a solution of potassium iodide in a test-tube, then a few drops of bromine water and shake. Can you explain the change that takes place?
(b) Add about 3 cm^3 of CCl_4 to an aqueous solution of potassium bromide in a test-tube, then add 3 cm^3 of chlorine water and shake. Is the result comparable with (a)?

4 Displacement of hydrogen halides from their salts

Attempt to displace the hydrogen halide from potassium chloride, bromide and iodide using a few drops of concentrated sulphuric acid. What happens? Are any other gases given off? Would it make any difference if syrupy phosphoric acid had been used? Try it.

5 Action of alkalis on halogens

(a) Add an equal volume of dilute sodium hydroxide solution to some saturated bromine water. The brown colour fades very quickly; what has been formed?
(b) Take some small lumps of iodine in a test-tube, and add 3 cm^3 of freshly made concentrated potassium hydroxide solution and warm. When the solution goes colourless after it has become hot add a little more iodine until it remains light brown in colour, then just decolourise this with a few drops of the alkali. Cork it and put the tube away to cool; crystals will form in the tube. What is the salt that is deposited?

6 Oxidising power of halogen oxyanions

(a) Oxidation of Pb^{2+} by ClO^-. See M4, Expt. 10.
(b) To two drops of ethanol in a test-tube add 5 cm^3 of sodium hydroxide solution and a large crystal of iodine; shake vigorously. The iodine dissolves forming IO^- which oxidises the ethanol to yellow crystals of iodoform, CHI_3.
(c) To 1 cm^3 of M/10 potassium iodide solution in a test-tube add 1 cm^3 of CCl_4, then run in a solution of M/100 potassium iodate

from a burette. The I^- is oxidised to iodine which dissolves in the organic layer to give a purple colour. Now to the whole of the liquid already present add an equal volume of concentrated hydrochloric acid, then some more potassium iodate solution with shaking. Describe the colour changes which take place. What is the oxidation state of the iodine present, and what is the equation for the overall reaction?

7   Oxidation of halide anions

(a) To small solid samples of potassium chloride, bromide and iodide in separate test-tubes add a small amount of solid manganese(IV) oxide. Mix well, then add a few drops of dilute sulphuric acid, warm and test for the gases given off. Are there significant differences between the various halides?

(b) Add a small amount of solutions of the three potassium halides to a dilute solution of copper(II) sulphate. Which halide ion is oxidised by Cu^{2+}? Can you account for the colour changes in this case?

A.7(8) THE TRANSITIONAL ELEMENTS

1   Formation of complexes

(a) Add dilute ammonium hydroxide solution drop by drop to a dilute solution of copper(II) sulphate. What is the formula of the intermediate precipitate? Is the deep blue $Cu(NH_3)_4^{2+}$ complex ion decomposed by (i) acids, (ii) H_2S?

(b) Repeat the experiment using silver(I) nitrate solution. Is this complex decomposed by (i) Cl^-, (ii) I^-, (iii) H_2S?

(c) Add dilute sodium thiosulphate solution drop by drop to a solution of silver nitrate until the precipitate at first formed redissolves. Boil the clear solution; what is the formula of the black precipitate formed? As before, test the stability of the complex to (i) Cl^-, (ii) I^-, (iii) H_2S.

(d) Add a few drops of CNS^- to a solution containing Fe^{3+}, and note the blood red colour of $Fe(CNS)^{2+}$. Now repeat the experiment in the presence of dilute sodium fluoride solution.

(e) Add concentrated hydrochloric acid drop by drop from a teat-pipette to a strong solution of pink cobalt(II) chloride, which contains the $Co(H_2O)_6^{2+}$ ion, and note the formation of $CoCl_4^{2-}$. Repeat the experiment stopping at the intermediate mauve stage; now warm this solution and note the change in colour.

2 Stabilisation of high oxidation states by complexing

(a) Mix equal volumes of 20 volume hydrogen peroxide solution and acidified cobalt(II) chloride solution: is there any change? Now add excess concentrated ammonia solution, and note the brown colour of the Co(III) complex formed.

(b) Cool some concentrated hydrochloric acid with lumps of ice in a conical flask, then add small quantities of solid manganese(IV) oxide from a spatula. Filter through glass wool, and note the brown colour of $MnCl_6{}^{2-}$, which decomposes readily even at 0°C.

3 Stabilisation of high oxidation states in alkali: reduction of high oxidation states in acid

(a) Mix equal volumes of 20 volume hydrogen peroxide solution and manganese(II) sulphate solution containing a little dilute sulphuric acid; nothing happens. Now make the solution alkaline with dilute sodium hydroxide solution. Note how the white manganese(II) hydroxide is oxidised to brown MnO(OH).

(b) Add 20 volume hydrogen peroxide solution to an alkaline solution of $CrO_4{}^{2-}$; the yellow colour persists. Now acidify with dilute hydrochloric acid and note the green colour of chromium(III).

(c) To a solution in which zinc is reacting vigorously with sodium hydroxide solution add a concentrated solution of ammonium vanadate, and note any colour changes which take place. Now repeat the reaction using zinc and sulphuric acid as the reducing mixture and compare the results.

(d) Shake a solution of $TiCl_4$ in strongly acidified zinc amalgam in a stoppered bottle. Note the colour changes which take place. Can you account for what has happened?

4 Polymerisation of oxyanions containing transitional elements

(a) To a concentrated solution of potassium chromate, add concentrated sulphuric acid drop by drop, until a precipitate is formed. This is $(CrO_3)_n$; filter the suspension through glass wool and examine the effect of one or two drops of ethanol on the dried precipitate (CARE).

(b) To 1 cm³ of dilute nitric acid saturated with ammonium molybdate solution, add a few drops of sodium orthophosphate solution and warm. A bright yellow precipitate is formed containing the ion $[PO_4(12MoO_3)]^{3-}$: all condensed phosphates give this test on boiling.

5 Catalysis

(a) Try the effect of a few grains of powdered MnO_2 on 3 cm³ of
100 volume hydrogen peroxide solution in a test-tube. The evolution
of oxygen gas is very striking, especially if a drop or two of liquid
detergent is included in the peroxide solution.

(b) To a solution containing 100 cm³ water, 2 cm³ dilute sulphuric
acid, 5 cm³ dilute potassium iodide solution, and a little freshly made
starch, add 1 cm³ of M/10 sodium thiosulphate solution from a
burette, then pipette into it exactly 1 cm³ of 20 volume hydrogen
peroxide solution. After a certain time interval the solution will go
blue, which corresponds to the liberation of free iodine. (If the
interval is too long or too short, adjust it by adding less or more
thiosulphate.) Compare this interval with that given by exactly
similar reaction mixtures to which one drop of solutions containing
(i) Ag^+, (ii) Cu^{2+}, (iii) MoO_4^{2-} have been added.

6 Formation of covalent compounds

(a) To a neutral solution containing Ni^{2+} add a few drops of a
dilute alcoholic solution of dimethylglyoxime,
$CH_3 . C(NOH) . C(NOH) . CH_3$, and filter off the scarlet precipitate
formed. Examine the stability of the complex (which is an internal
salt containing two molecules of the anion of the complexing agent)
in acid and alkali, then test its solubility in ethanol and diethyl ether.

(b) Intimately mix some powdered sodium chloride and some
powdered potassium dichromate, transfer to an ignition-tube and
add cautiously some concentrated sulphuric acid and warm gently. An
orange vapour is formed which if poured into a solution of lead(II)
nitrate will give a yellow precipitate of $PbCrO_4$, thus demonstrating
that the vapour contains a compound of Cr(VI). Suggest a possible
formula.

Introduction to Part B

The rise of the concept of orbitals

1 THE STRUCTURE OF THE ATOM IN TERMS OF ATOMIC NUMBER

Atoms are composed of protons, neutrons and electrons. Although the discovery of these particles is a fascinating story in physics, the chemist is more concerned with the way that these particles are arranged within the atom.

Before the properties of these particles were fully elucidated, a brilliant piece of work by Moseley (working under Rutherford in Manchester in 1913) had given a theoretical, rather than an empirical, basis to the order of elements in the Periodic Table. When any element is bombarded with streams of fast-moving electrons in a high vacuum, X-rays are emitted. Like any electromagnetic waves, X-rays can be diffracted through a crystal grating to give a spectrum which can be photographed. Successive elements showed a regular shift in the wave length of their X-ray spectra as the total number of electrons stripped off from the atom of the element increases.

It had always seemed peculiar that ragged increases in atomic weight were accompanied by an orderly gradation of properties. It is much more satisfying to appreciate that the change in chemical properties from sodium, element 11 in the Periodic Table, to magnesium, element 12, corresponds to a gain of one electron and the corresponding proton, rather than to an increase of 1·32 units of mass.

This discovery provided an immediate explanation for the existence of isotopes, first found for the element neon by Thomson and Aston in 1919. Isotopes contain the same number of protons and electrons but the number of neutrons in the nucleus varies. Isotopes have identical chemical properties, strong evidence that the latter depend upon the number of extra-nuclear electrons.

2 EXPERIMENTAL DEMONSTRATION OF THE EXISTENCE OF THE NUCLEUS

A very simple experiment shows that the neutrons and protons are gathered into a compact nucleus with a radius some 10^5 times smaller than the distance of closest approach of two atoms in a di-atomic molecule. Rutherford and his pupils bombarded very thin metal foils with alpha particles (helium ions) obtained from the decay of radioactive elements, and were astonished to find that while the vast majority of the alpha particles went straight through the foil, some were deflected over a range of angles which in a very few cases was actually 180°—that is to say, the alpha particles were reflected straight back towards the source. A good analogy is the deflection of rifle bullets by the struts of an electric pylon: most bullets pass through the pylon, a few will glance off the struts, and very rarely one will ricochet straight back. In the words of Rutherford himself "it was about as incredible as if you had fired a fifteen inch shell at a piece of tissue paper and it came back and hit you". Clearly a tiny central nucleus will have less to do with the interactions of one atom with another than the number and disposition of the electrons at its periphery.

It would be highly satisfactory if some experimental method could be discovered for measuring the position and trajectory of the extra-nuclear electrons in any atom, but so far no technique has been found. This had led to the construction of a hypothetical system known as quantum mechanics, whose postulates are directly contrary to 'common sense' derived from experience in the every day world. In this system the possibility of the electron having a trajectory is flatly denied. This is just as well, for any charged particle accelerated in a path around the nucleus should radiate energy and collapse into the centre of the atom—leading to the destruction of the universe as we know it in a fraction of a second.

3 FUNDAMENTAL POSTULATES OF QUANTUM MECHANICS

The three fundamental postulates of quantum mechanics are as follows. First, that the electrons in an atom exist in certain states of fixed energy, and in accordance with Planck's quantum theory transitions can occur between these states by the absorption or emission of energy usually in the form of light: second, that matter has both a particle and a wave aspect: thirdly, that it is never possible to know precisely the position and momentum of any particle simultaneously. All these postulates can be stated in mathematical form:

(i)	$$\Delta E = h\nu$$	E = energy ν = frequency λ = wavelength
(ii)	$$\lambda = \frac{h}{mv}$$	m = mass v = velocity
(iii)	$$\Delta(mv)\Delta x = h/4\pi$$	x = distance

In all these equations h is Planck's constant, the ultimate basis for all the discontinuity in our universe.

There is some experimental evidence to support these postulates in the sub-microscopic world. First, the emission spectrum of hydrogen is a fine demonstration of the reality of finite discontinuous energy states. Secondly the electron can indeed manifest both a particle and a wave nature; the photo-electric effect depends upon considering it as a particle, but its diffraction in the electron microscope can only be explained on a wave basis. Thirdly, a particle occupies a given point in space whereas a wave takes time to pass through it; these two points of view can only be reconciled by admitting a range of uncertainty into the simultaneous determination of quantities concerned with space and quantities concerned with time: this is known as Heisenberg's Principle of Indeterminacy. Experimentally the broadening of spectral lines (denoting an increasing indeterminacy in energy) can be observed for transitions for states of very short life whose indeterminacy in time is therefore necessarily small. Extraordinary though its postulates may seem, quantum mechanics is retained for the same reason as any other hypothesis—they unify a great deal of previously unconnected data. Unhappily the predictive power of quantum mechanics is severely limited by the mathematical complexity of the solutions involved.

4 ELECTRON ENERGIES AND SPECTRA

If questions about the position and trajectory of an electron in the atom are not allowed because they are meaningless in terms of the fundamental postulates, what is there left? Only the energy of the extra-nuclear electrons in states of long life remains, deduced from the frequency of lines in atomic spectra. The hydrogen atom is the simplest possible example and repays detailed study.

When the light from an incandescent object is shone through a monatomic vapour under the correct experimental conditions, and then passed through a prism, dark lines are observed at specific

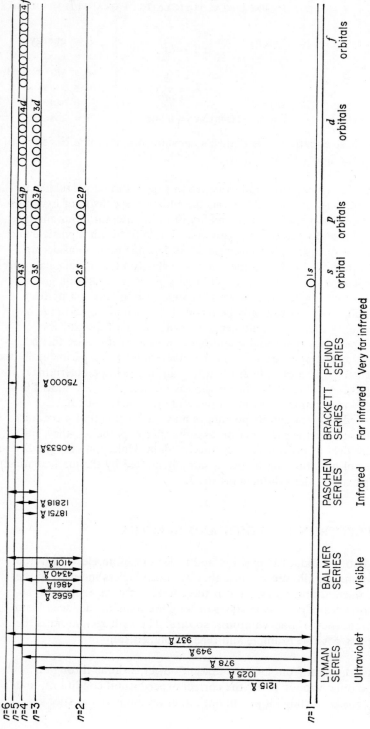

Fig. 3. Relationship between the atomic spectra of hydrogen and electronic energy levels.

wavelengths on a continuous light background: this is called an **absorption line spectrum,** since light is being absorbed by the atom. Similarly if an electric discharge is passed through gases at low pressure they tend to glow, and if the light from this glow is passed through a prism it generates sharp bright lines on a continuous dark background, the lines being in exactly the same positions as in the absorption spectrum. These **emission spectra** can also be obtained from gaseous atoms in hot Bunsen flames—e.g. the characteristic flame colours given by the alkali metals.

Consider as an example the emission spectrum of hydrogen. There are several series of distinct spectral lines; one in the ultraviolet part of the spectrum, one in the visible and at least two in the infrared. The ultraviolet lines naturally correspond to the largest energy changes, and as the frequency increases they draw closer and closer together, finally becoming continuous at a frequency of approximately 3.2×10^{15} cycles per second, corresponding to an energy of 217.9×10^{-20} joules.

The frequencies of these converging lines offer an irresistible temptation to a mathematician, who will long to find a simple formula to express their magnitude and spacing. The formula for the energies of the successive ultraviolet spectral lines of hydrogen is very simple and was first recognised by a Swiss schoolmaster called Balmer. It is $217.9 \times 10^{-20}/(n_i)^2 - 217.9 \times 10^{-20}/(n_{ii})^2$, where n_i is unity and n_{ii} can have any integral values. This expression reaches its maximum when n_{ii} is infinite, though it is difficult to observe the precise point at which the spectrum becomes continuous. More gratifying still is the fact that the wavelength where the visible and the two infrared spectral series become continuous follow from putting n_i equal to 2, 3 and 4, and once again letting n_{ii} become infinite. In these expressions n_i is called the **principal quantum number.**

With the aid of Figure 3 the significance of the results is easy to see: there are levels of increasing energy corresponding to increasingly loose binding of the electron to the nucleus, and transitions between them are possible. The **ground state** corresponds to the tightest binding of the electron to the nucleus and the minimum energy of the system. This value will be large and negative in sign, since the value of zero has been arbitrarily allotted to the energy of a proton and electron at infinite separation. If the proton and electron are closer than this, the potential energy has been lowered and energy evolved— hence the negative sign. The differences in the energies of the available quantum states become smaller as the electron is more loosely bound. Quantitatively the ground state energy will be —217·9 × 10^{-20} joules, and the limit beyond which the spectrum is continuous corresponds to the electron gaining enough energy to leave the proton entirely.

The reason why all the lines of the hydrogen spectrum can be characterised by just a few integers (quantum numbers) is a direct consequence of the wave nature of the electron. An everyday analogy will make this clear. When standing waves are generated in a rope by shaking the end rhythmically, the wave pattern always fits neatly into the space available and there are an integral number of points where there is no motion. The distance between two such points is half a wavelength, and the pattern does not change with time. The wavelength is given by (length of string)/n; and the length of the string is referred to as a *boundary condition*. In the case of the hydrogen atom, the system can accept standing electron waves only of fixed wavelength, the boundary condition here being the ground state energy of the system. The integer necessary to define the wavelength of the emitted light in terms of the boundary conditions is n_{ii} in the Balmer formula.

5 DISTRIBUTION OF ELECTRONS WITHIN THE ATOM

Since the energy of the stationary states which can be occupied by the electron in a hydrogen atom is very accurately known, on Heisenberg's Principle of Indeterminacy its position cannot be determined with any precision. In fact the only meaningful way to express this position is in terms of a *probability distribution*. The word **orbital** is often used to describe this probability distribution. An orbital description of the motion of an electron contains the same sort of information conveyed by holes made by darts in a dartboard. After the board has been used in many games the distribution of holes shows how successful earlier players have been in their scoring; the 20, the 19 and the bull's-eye will be heavily peppered, and the smaller numbers will register correspondingly fewer hits. The holes in the dartboard tell only the probability that a given throw will land in a particular place. It does not tell us the order in which the holes were made in the dartboard. In the case of the electron probability distribution, the orbital gives the probability that an experiment designed to locate the electron will find it a particular distance from the nucleus; but it does not tell how the electron moves from point to point. A two dimensional representation of a three dimensional probability distribution is given below in Figure 4.

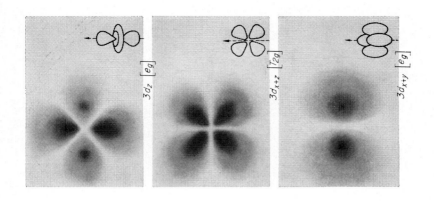

Scale of densities

Probability of finding electron in one cubic Ångstrom:

1 chance in 2, or better:

1 chance in 10:

100:

1,000:

10,000:

100,000:

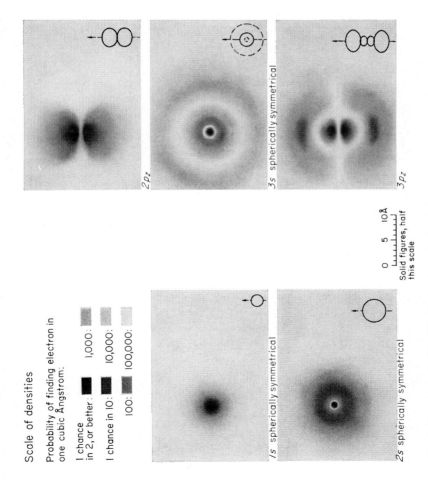

1s spherically symmetrical

2s spherically symmetrical

2p_z

3s spherically symmetrical

3p_z

$3d_z$ $[e_g]$

$3d_{x+z}$ $[t_{2g}]$

$3d_{x+y}$ $[e_g]$

0   5   10 Å

Solid figures, half this scale

Fig. 4. Electron probability distribution in the hydrogen atom.

In probability distributions, where the outermost electron may be a long distance from the proton or protons, there are only two meaningful radii. First that at which the electron density—number of electrons per unit volume—is largest, and secondly that radius outside which there are as many electrons as there are inside. Both these radii can be derived from the probability distribution, so they can be quoted exactly without infringing the Principle of Indeterminacy, but there is no boundary radius outside which there is NO chance of finding the electron; after all some dart players hit the floor rather than the board!

In quantum mechanics the distinction between potential and kinetic energy still holds, but the visualisation of the latter is difficult. On consideration, it makes sense to identify kinetic energy with the energy needed to compress the electron into a confined space. Now for any coulombic system the Virial theorem states that the kinetic energy is equal to half the potential energy with the sign reversed. Thus the lowering of potential energy which follows when electrons are attracted to more than one nucleus is the dominant factor in bond formation, but this is always offset by a rise in kinetic energy corresponding to the increased compression of the electrons between the two nuclei.

6   THE STABILITY OF THE NOBLE GAS CONFIGURATION

Since any attempt to postulate electron trajectories is meaningless, an approach to atomic structure through energy relationships seems obligatory. Once more this is where empirical knowledge can help tremendously. The noble gases have quite exceptional chemical stability, which is presumably connected with the extreme difficulty they have in gaining or losing a peripheral electron. This leads immediately to the conclusion that the particular distribution of the electron population associated with the noble gases possesses a special stability, so much so that neighbouring atoms will tend to adopt the noble gas configuration by gaining or losing electrons.

Regularity among the electron populations of the noble gases

Noble Gas	Electrons	Differences
Helium	2	2–0 = 2
Neon	10	10–2 = 8
Argon	18	18–10 = 8
Krypton	36	36–18 = 18
Xenon	54	54–36 = 18
Radon	86	86–54 = 32

Thus O^{2-}, F^-, Ne, Na^+, Mg^{2+} are isoelectronic, all having ten electrons but an increasing number of protons. The regularity of the electronic structure of the noble gases is best shown by a table (p. 81). These results are extremely suggestive, as the following exercises show.

(a) Consider the series 2–8–18–32 (forgetting for a moment that 8 and 18 occur twice in the series). These numbers were obtained by taking differences. Take differences again and use them to predict the next number above 32 in the series above.
(b) Divide the numbers 2–8–18–32 by two. Use these numbers as a basis for predicting the next number after 32 in the series by another method rather than taking differences.

7 DISTRIBUTION OF ORBITALS ASSOCIATED WITH A PARTICULAR QUANTUM NUMBER

In the hydrogen atom, the principal quantum number n can be shown to have associated with it n^2 orbitals, with identical energies, but different spatial probability distributions, by some fairly simple quantum mechanical calculations. Thus when n is 1 there is only one distinct orbital, when n is 2 there are four distinct orbitals, when n is 3, nine distinct orbitals, when n is 4, sixteen distinct orbitals. If the extra assumption is made that two and only two electrons can occupy a given orbital (a purely empirical assumption known as the Pauli Principle) the number of electrons in successive orbitals becomes 2–8–18–32, in perfect agreement with the result for the differences between the electron populations of the noble gases in the last section.

The lowest energy level of a hydrogen atom, when n equals 1, corresponds to an electron distribution that is calculated to be spherically symmetrical about the nucleus (see Figure 4): this is called the $1s$ orbital. When n is 2 there are four orbitals; again one of these is spherically symmetrical, but there are three further orbitals without spherical symmetry. Their calculated shapes are dumb-bells along each of the three cartesian axes perpendicular to each other; thus the p_z orbital is concentrated in the z direction and so the p_z electron is more likely to be found near the z axis than anywhere else (Figure 4). The directional character of these probability clouds is the basis for theories about the geometry of molecules. Every energy level with a value of n greater than 1 has three p orbitals of roughly similar shape, as well as a spherically symmetrical s orbital.

For *n* equals 3, there are according to the calculations nine possible
orbitals. Four of these are accounted for by the 3*s* and 3*p* orbitals;
the remaining five orbitals are called 3*d* orbitals, and their shapes are
more complex. The pattern repeats itself for *n* equals 4 so beside the
single 4*s*, three 4*p* and five 4*d* orbitals there are seven extra orbitals,
the 4*f* orbitals, which are of even more complex shape (see Figure 4).

8 MODIFICATION OF HYDROGEN-LIKE SPECTRA FOR MORE COMPLEX ATOMS

All atoms display line spectra, but in general these spectra are much
more complex than that of atomic hydrogen. Nevertheless the
similarities naturally lead to the supposition that all atoms possess
only particular energy levels; once again spectral lines correspond to

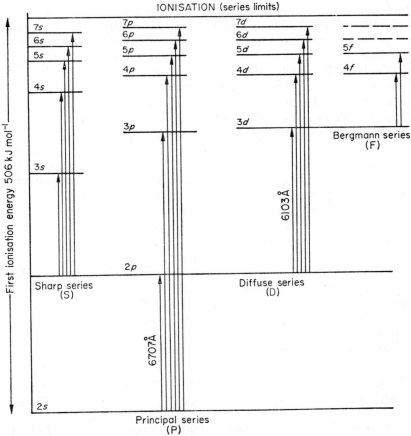

Fig. 5. Energy level diagram of the lithium atom.

transitions between energy levels. Figure 5, showing how lines in the lithium spectrum can be interpreted in terms of energy levels is very instructive. Notice the two vital differences of this spectrum compared with that of hydrogen: not all the orbitals associated with a particular principal quantum number have the same energy, and not all transitions between energy levels are allowed. Such a diagram is purely experimental; the mathematics involved in attempting to compute it from quantum mechanics are quite impracticable at present, even with the largest computer.

9 SCREENING OF EXTRA-NUCLEAR ELECTRONS BY FILLED ORBITALS

How would the energy of an orbital associated with a particular principal quantum number be expected to change with an increase of nuclear charge? Helium, with two protons and two electrons in the $1s$ orbital, would be expected to require, by Coulomb's Law, double the energy to remove the first electron (less a small amount to mutual electron repulsion) than would be required to remove the single electron from hydrogen. Such is found to be the case: to remove an electron from the $1s$ orbital of hydrogen requires 1300·8 kJ mol^{-1}, while to remove one from the $1s$ orbital of helium requires 2370 kJ mol^{-1}.

The amount of energy required to remove an electron from an atom in the gas phase is known as the **ionisation energy.** It can be measured comparatively readily for any atom by bombarding it with projectiles (usually electrons) of steadily increasing kinetic energy, and detecting the critical value at which positive ions are formed by the target. If the kinetic energy of the projectiles is increased, a second electron will be removed from the element which gives a corresponding value for a second ionisation energy, and the process may be repeated. Thus for aluminium the first four ionisation energies are: 577, 1820, 2740 and 11600 kJ mol^{-1}. The large discontinuity between the third and fourth ionisation energies of aluminium shows that the fourth electron has been removed from the stable orbitals corresponding to the noble gas structure of neon.

Now consider the case of lithium, with three electrons, the third of which is a $2s$ electron. The energy required to remove an electron from the $2s$ orbital of hydrogen is, from the Balmer formula, 1310/4 or 328 kJ mol^{-1}, while for lithium the removal of a $2s$ electron requires 521 kJ mol^{-1}. This can be explained in the following way.

Lithium has a charge of $+3$; the two $1s$ electrons presumably shield the $2s$ electron from two of the positive charges, since the main volume of probability of the $2s$ electron lies largely outside that of the two $1s$ electrons. But there is a small probability that the $2s$ electron will be found close to the nucleus (see Figure 4), and here the shielding effect of the $1s$ electrons is lost and the $2s$ electron is acted

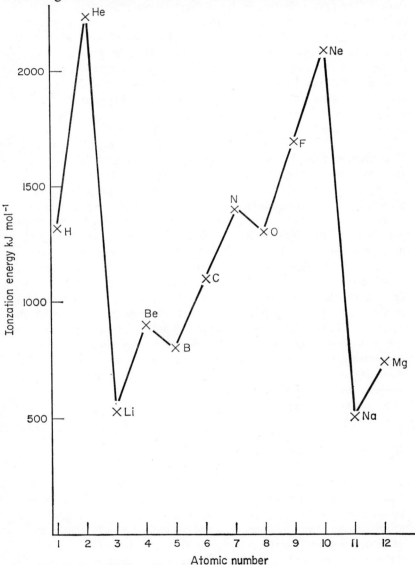

Fig. 6. First ionisation energies of the first twelve elements.

on by the full charge of $+3$, and so it is more firmly bound than that of hydrogen. The plot of ionisation energy against atomic number is very instructive for the second period. Ionisation energy reaches a local maximum at atomic numbers corresponding to the noble gases, whereas the succeeding alkali metals show a local minimum; thereafter the ionisation energy increases irregularly across the Period. The first p electron in boron has a lower ionisation energy than the second s electron in beryllium (see Figure 6). Presumably the first $2p$ electron, the orbital occupied by which does not penetrate through the $2s$ electron cloud, feels an attractive charge of only $+1$ owing to the shielding effect of the filled $2s$ orbital, whereas the second $2s$ electron feels the effect of a net charge of $+2$. In accordance with Hund's Rule (p. 88) a slight fall between the third and fourth p electrons corresponds to the electronic repulsion involved in the first double occupancy of a $2p$ orbital.

Each element therefore must possess a unique energy level diagram. However, the features that appear in the lithium diagram are fairly general and worth detailed comment. Firstly, np orbitals are always higher in energy than ns orbitals, a fact simply explained in terms of the probability distribution of np orbitals lying for the most part further from the nucleus than that of ns orbitals, so that the latter will effectively screen the former. Secondly, nd orbitals are higher in energy than $(n+1)s$ orbitals. In contrast to the ns orbitals, which all penetrate close to the nucleus, and the slightly penetrating np orbitals, the probability of finding an nd electron near the nucleus is very small. The main probability for the nd electrons lies nearer the periphery of the atom than that of the ns and np electrons, so in the potassium atom (but not in the titanium atom, see p. 143) the penetrating $4s$ orbital is more stable than the $3d$ orbital. Thirdly, the same argument applies with even more force to the non-penetrating nf electrons, as compared to the nd, np and ns electrons: thus nf electrons are of even higher energy than nd electrons.

10 FILLING THE AVAILABLE ORBITALS

The development of the entire Periodic Table should now be clear. The special stabilities of the noble gases are fixed by the large gaps between the energy levels (Figure 7). The number of orbitals of approximately the same energy is not controlled by n, the principal quantum number, for the energies of ns, np, nd and nf electrons differ widely owing to the effect of screening. Instead of using the clumsy *number of orbitals of approximately the same energy* the phrase **rhythmic electron pattern** will be used. This is preferable to the term *electronic subshell,* for the latter suggests a static and non-penetrating set of electrons which is far from the truth. The number of orbitals in a particular rhythmic electron pattern, doubled on account of the Pauli Principle which allows two electrons to occupy each orbital, corresponds with the number of electrons between one noble gas and the next. The existence and composition of these rhythmic electron patterns in many electron atoms is deduced from the empirical sequence 2–8–8–18–18–32, the electron intervals which separate the noble gases, supported by a detailed knowledge of

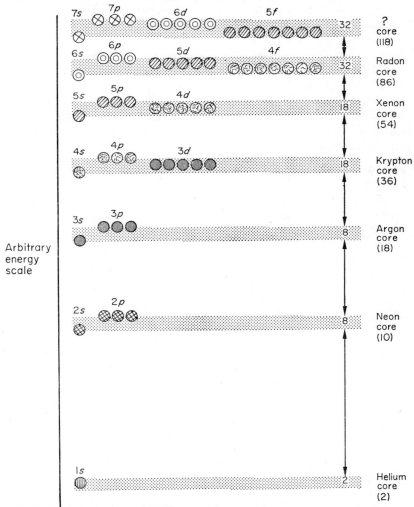

Fig. 7. A schematic energy level diagram of a many-electron atom.

atomic line spectra. Nevertheless it is essential to keep in mind the certainty that this sequence could *not* have been deduced from first principles; once again it is the interplay between theory and experiment that makes science a going concern.

On passing from one element to the next in the Periodic Table, the available orbital of lowest energy is always preferred for the incoming electron: the extra negative charge is offset by the accommodation of an extra proton (and probably some extra neutrons) in the nucleus. The presence of an extra proton lowers the energy of all the filled orbitals and causes them to contract in towards the nucleus, but their energies relative to each other are usually unchanged.

One extra assumption is necessary in order to define the electron population of every atom; this is Hund's Rule of Maximum Multiplicity. This rule recognises the mutual repulsion of electrons by stating that no electron would join another in a single orbital if a vacancy in an empty orbital of the same energy is available. This can be illustrated by writing a few typical electron configurations for certain elements.

Helium $\quad$ $1s^2$

Lithium $\quad$ $1s^2.2s^1$

Nitrogen $\quad$ $1s^2.2s^2.2p_x^1.2p_y^1.2p_z^1$ $\quad$ (not $2p_x^2.2p_y^1$)

Neon $\quad$ $1s^2.2s^2.2p^6$ $\quad$ (no need to specify $2p_x^2.2p_y^2.2p_z^2$)

Scandium $\quad$ $1s^2.2s^2.2p^6.3s^2.3p^6.4s^2.3d^1$

Part B: Theoretical Interpretation

B.1
The modern Periodic Law

B.1(1) STATEMENT OF THE LAW IN THEORETICAL TERMS

Mendeleev's Periodic Law can be restated in electronic terms:
"Since the electronic structure of the atoms of the elements vary
periodically with atomic number, all properties dependent on this
structure tend also to vary periodically with atomic number."

B.1(2) THE ENLARGED SCOPE OF THE RESTATED LAW

The new Law is stated in more abstract terms than the old, but the
whole idea that rhythmic electronic patterns are periodic depends on
the earlier idea of chemical periodicity. Enlarged hypotheses are not
daringly wider generalisations than those they supersede, for if so
they would be all the more liable to upset by new experimental data.
The relationship between the two is this; the terms in which the new
hypothesis is stated (in this case the idea of periodicity) is only
comprehensible in the light of the earlier work.

The enlarged scope of the Law as restated makes the apparent
exceptions of section A.2(4) immediately comprehensible.

(a) The revised classification of elements as Main Group,
transitional, and inner transitional removes the necessity of stressing
similarities between Mendeleev's A Group and B Group elements.
(b) The realisation that atomic weight is not a fundamental property,
but only a statistical average of the available isotopes, makes the
tellurium/iodine exchange comprehensible. Iodine has but one stable
isotope of mass 127, whereas tellurium has several, ranging in mass
from 124 to 130, the latter having an abundance of 34%. By
multiplying each mass by its abundance and dividing by 100, a value
of 127·6 is obtained for the atomic weight of tellurium.
(c) The transitional elements in the new periodic arrangement
extend from scandium to copper for the first transitional Period, so
there is no arbitrary separation of iron, cobalt and nickel from other
similar elements.
(d) The discovery of all the rare earth elements has made their
classification as an inner transitional series plausible, especially as the
actinons (elements following actinium) resemble the rare earths in the
same kind of way as the second transitional Period resembles the first.

B.1(3) PREDICTIVE POWER OF THE MODERN PERIODIC LAW

(a) The trans-uranic elements were synthesised by the bombardment of existing nuclei with particles of carefully moderated energy, so that in all cases the element became available before the chemistry of its compounds was known. Predictions about the nature of the compounds formed and the possible oxidation states of these elements were justified to a remarkable degree by experimental work using capillary tubing and carrier materials. Particularly noteworthy was the prediction of the stable $+3$ oxidation state of curium (whose $5f^7$ electron configuration makes it similar to gadolinium) and successive elements.

(b) Once the first metallic carbonyl, $Ni(CO)_4$, had been isolated, the formulae of other metallic carbonyls like $Cr(CO)_6$, $Fe(CO)_5$, $Co_2(CO)_8$ could be predicted and were later found.

(c) Once compounds of the noble gas xenon had been isolated, the periodicity of ionisation energy demanded that krypton be the next noble gas to be investigated: KrF_2 was the result.

(d) Once the first 'sandwich compound', dicyclopentadienyl iron $Fe(C_5H_5)_2$ was isolated, it was clear that other compounds which involved the co-ordination of the aromatic sextet might also exist: di-benzene chromium, $Cr(C_6H_6)_2$ is one of the hundreds that have so far been found.

In the latter three cases the initial breakthrough was an empirical one, but, once made, the path to follow was indicated by the Periodic Law.

B.1(4) ANOMALIES IN THE MODERN PERIODIC LAW

(a) The reluctance of the two $6s$ electrons to ionise is used as an explanation (the 'inert pair' effect) for the anomalous chemistry of mercury and for the oxidation state shown by neighbouring elements being two less than their maximum. There is no theoretical justification for these empirical observations; here is a clear case of giving something a name to render it less mysterious!

(b) Gold, for no predictable reason, shows an oxidation number of $+3$ in a large number of complexes.

(c) Chromium and cobalt form an enormous number of co-ordination complexes, the majority being with nitrogen-containing compounds. Manganese and iron lie between them; but they form fewer complexes and the majority seem to be with oxygen-containing compounds. This special affinity of one element for another is notable in other parts of the Periodic Table (Hg-N is an outstanding example), but so far it is entirely unexplained.

B.2
Theoretical definitions

B.2(1) MAIN GROUP ELEMENTS

Those elements which either gain or lose electrons from s and p orbitals only to form simple ions.

B.2(2) TRANSITIONAL ELEMENTS

Form some compounds in which there is an incomplete rhythmic pattern of d electrons.

B.2(3) INNER TRANSITIONAL ELEMENTS

Form some compounds in which there is an incomplete rhythmic pattern of f electrons.

B.2(4) ELECTROPOSITIVITY

The **Standard Electrode Potential** is a quantitative measure of the ease with which a metallic element is deposited at the cathode. To construct a scale, it is necessary to have a reference point, and hydrogen is arbitrarily given a value of zero. Using the IUPAC sign convention, lithium, the most electropositive element, has a potential of $-3 \cdot 08$ volts under standard conditions. (The negative charge for the most electropositive metal looks unfortunate at first sight. However, if the tendency to lose ions to the solution exceeds the tendency for ions to deposit on the metal, as it does in lithium, the residual electrons that remain on the metal will give it a negative charge.)

The standard electrode potential of any other element is measured against a standard hydrogen electrode (or some more convenient reference standard) by setting up a cell of the type

$$\text{Pt, H}_2 \mid \text{H}^+ \text{ (molar)} \parallel \text{M}^+ \text{ (molar)} \mid \text{M (standard state)}$$
$$\text{salt bridge}$$

and measuring the voltage in the external circuit with a valve voltmeter. The cathode is that electrode which receives electrons from the external circuit.

THREE STAGES ARE INVOLVED: M (metal) $\xrightarrow{\text{atomisation}}$ M (gaseous atoms)

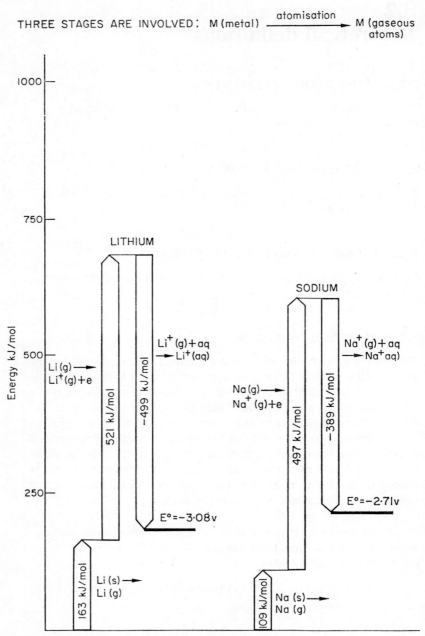

Fig. 8. The conversion of a metal atom into its hydrated ion.

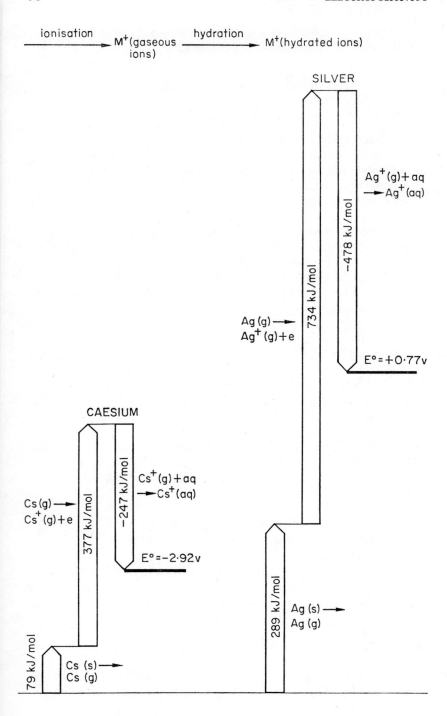

Three factors, all involving large energy changes, contribute to the relative electrode potential of any element.

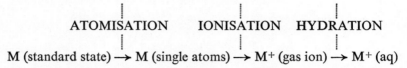

$$\text{M (standard state)} \longrightarrow \text{M (single atoms)} \longrightarrow \text{M}^+ \text{ (gas ion)} \longrightarrow \text{M}^+ \text{ (aq)}$$

By subtracting the heat of hydration estimated for the single metal ion from the sum of the atomisation and ionisation energies, the approximate relative value of the electrode potential can be calculated; the results for some elements in Main Group I (with silver for comparison) are given in Figure 8.

Any figure based on the difference between large quantities, as is the standard electrode potential, tends to be somewhat irregular and too much emphasis should not be put on small variations. The general trend in the Periodic Table is clear: the relative electrode potential falls along a period but increases down a Group, since in most cases it is the ionisation energy which is the controlling factor (see p. 84).

B.2(5) VALENCY

The word valency can have a large number of different meanings depending upon the context. It can refer to the charge on an ion, the oxidation number, the total number of atoms to which a particular atom will be bound, or the number of hydrogen atoms which one atom will combine with or release in a chemical reaction. The phrase "Valency is a small whole number" covers a multitude of sins of omission. It is probably better to jettison a word so debased, so it is used in this book only as an adjective.

Valence electrons are those electrons in an atom that are loosely enough bound to take part in compound formation.

Valence orbitals are the entire rhythmic pattern of orbitals, filled or unfilled, of about the same energy as those which are occupied by the valence electrons.

B.2(6a) IONIC BONDING

Ionic bonding is one of the three idealised mechanisms for the formation of a chemical bond; the other two are covalent bonding

and metallic bonding. All real bonds are a combination of the
properties of the three idealised types. A chemical bond is said to
exist when two adjacent atoms are so located that the potential
energy of the system is a minimum, and when appreciable energy
must be supplied to cause the atoms to separate. In general, ALL
BONDS FORM BECAUSE ELECTRONS ARE
SIMULTANEOUSLY ATTRACTED TO TWO OR MORE
NUCLEI, thus lowering the potential energy of the system.

The best way to understand ionic bonding is to take a specific
example and consider each step in the process of converting metallic
lithium and gaseous fluorine into an ionic crystal of lithium fluoride.
The stages involved are given in the Born-Haber cycle diagram
overleaf (Figure 9).

It is immediately clear that the formation of single gaseous ions of
lithium and fluorine is a process that takes in a great deal of energy,
414 kJ mol^{-1}. So much for the often quoted catch phrase that the
driving force in the formation of electrovalent compounds is the
formation of ions: rather it is the enormous fall in potential energy
when the separate ions pack tightly together to form a crystal lattice.
The lattice of lithium fluoride is probably best described as a three
dimensional co-ordination polymer, in which each of the six fluorine
ions equidistant from the lithium ion supplies alternately a pair of
electrons to fill the four empty orbitals round the metal ion.

Another widely quoted fallacy is that electrovalency is "the complete
gain or loss of an electron". It is a remarkable fact that the volume
of one mole of lithium atoms in the metal is actually greater than the
volume of one mole of lithium ion and one mole of fluorine ions
combined together in lithium fluoride. So the electron which the
lithium atom is supposed to have 'lost' to the fluorine atom is
actually closer to the lithium nucleus in lithium fluoride than it is in
metallic lithium!

Main Group elements never form more than one single ion. The
theoretical maximum number of electrons which such elements can
lose or gain in ionic bonding is fixed by the arrangement of filled
orbitals in the noble gasses next above and next below them. For
instance Mg^{2+} could never lose a further electron in a chemical
reaction to form Mg^{3+}, since the third electron would need to be
drawn from the deeply buried $2p$ orbitals against the attractive force
of no less than ten protons.

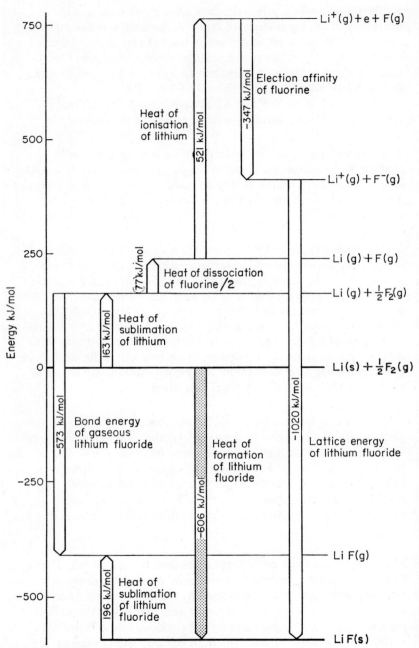

Fig. 9. Born-Haber cycle for lithium fluoride.

The argument which establishes why Mg^+ is not formed is a little more subtle. If a Born-Haber cycle is constructed for the theoretical compound MgCl, using the experimental lattice energy of NaCl as a rough approximation, the heat of formation obtained for MgCl is indeed exothermic, but nothing like so energetically favourable as the -642 kJ mol^{-1} which is the experimental heat of formation of $MgCl_2$. This is because the much greater lattice energy of $MgCl_2$ as compared to MgCl more than outweighs the extra ionisation energy required to convert Mg to Mg^{2+} rather than Mg^+, and to ionise an extra $\frac{1}{2}Cl_2$.

Since simple ions of very high charge are never formed, transitional metal ions do not possess noble gas electron configurations. The energy barriers to the removal of successive electrons from these elements are not large, so several ions may be formed, e.g. $Fe(H_2O)_6^{2+}$ and $Fe(H_2O)_6^{3+}$; very often such ions may be interconverted by mild oxidising and reducing agents.

B.2(6b) POLARISATION

Cations are normally small and compact compared to anions, since there is an excess of protons to attract the electron cloud in the former and a deficiency of protons in the latter. The smaller a cation and the higher its charge, the more readily will it deform the spherical electron cloud in an anion, and drag it into the region between the two nuclei. This strengthens the attraction between the ions; indeed it is the first step towards true covalent bonding. A rough guide to the polarising power of cations is given by the ratio charge/radius; for the alkali metals it is less than 1, for lithium $1\frac{1}{2}$, for calcium 2, for beryllium 7. Fajans' Rules indicate the conditions that favour minimum polarisation: large cations of low charge minimise the charge/radius factor, while the electron clouds of small anions are much more difficult to deform than those of large anions—e.g. the polarisability of I^- is greater than that of F^-. Compact symmetrical oxyanions like ClO_4^- and SO_4^{2-} are also very difficult to polarise.

B.2(7a) COVALENT BONDING

The mechanism of covalent bonding is best expressed in terms of molecular orbitals, formed by the overlap and rearrangement of atomic orbitals. Two atomic orbitals from different nuclei always overlap to form two molecular orbitals, one of lower energy than the atomic orbitals and one of higher energy. A crude visual analogy for this energy split is that the wave functions of the overlapping atomic orbitals may be either in or out of phase.

The total electronic structure of any compound is built up by successively feeding electrons into the orbital of lowest energy which is vacant. The lower energy **bonding orbitals** are filled first, but if these have no vacancy available the electrons must find a place in the higher energy **anti-bonding orbitals**. The *Bond Order* of a particular bond is given by:

$$\frac{\text{Number of electrons in bonding orbitals} - \text{number of electrons in anti-bonding orbitals}}{2}$$

Bonds with an order of 1 (the electron pair bond first postulated by Lewis in 1916) are extremely important, but there is nothing sacred about the theory that a pair of electrons are necessary before bonding can take place. H_2^+ and He_2^+, both of which are formed in spark discharge tubes, have binding energies of over 250 kJ mol^{-1}; in the first case there is one electron in a bonding molecular orbital, in the second there are two electrons in a bonding molecular orbital and one in an anti-bonding orbital; thus the bond order for both is $\frac{1}{2}$.

The prevalence of the two electron bond has its origin in (a) the Pauli Principle, which forbids three electrons to occupy the same region of space at the same time and (b) because a two electron bond always has a bond energy approximately double that of a one electron bond. However fractional bond orders do occur in stable molecules and ions (cf. the bond order of $2\frac{1}{2}$ for NO, $1\frac{1}{2}$ for O_2^-), and in certain cases, e.g. O_2, two one-electron bonds are preferred to one two-electron bond, for in this way inter-electronic repulsion can be minimised.

B.2(7b) ELECTRONEGATIVITY

Just as the concept of polarisation helps to illumine the gradual changes from almost pure ionic bonding towards covalent bonding, so does the concept of electronegativity cast light upon the reverse process by focusing attention on the unequal sharing of electrons between the atoms in a covalent bond. There is no fully agreed way of making quantitative calculations, but a qualitative knowledge of the excess (δ^+) or deficiency (δ^-) of charge on a particular atom in a compound is very valuable in predicting the reactions of this atom.

One possible quantitative approach is to calculate the force on the electron at the covalent radius of the atom (taken to be half the distance of a single bond between two of the atoms) allowing for the shielding effect of filled atomic orbitals, then using heats of reaction as a guide, to convert these values to an arbitrary scale of electronegativity running from Li at $1 \cdot 0$ to F at $4 \cdot 0$. On this scale H is $2 \cdot 1$ and C is $2 \cdot 5$; thus in a bond between them carbon has a partial negative charge and hydrogen a partial positive charge. This accounts for the inductive effect in organic chemistry, a very powerful concept for elucidating structure and mechanism. For instance CH_3—$COOH$ is a weaker acid than CCl_3—$COOH$, for in the former there is no withdrawal of electrons from the COOH group to help it to release a proton.

There are two further examples of the use of the concept that are very striking. First, the stability of bonds between identical atoms is lowered if they both carry the same partial charge. Thus H_2N—NH_2 is rather unstable because both nitrogen atoms carry a partial negative charge, whereas H_3Si—SiH_3 is rather unstable because both silicon atoms carry a partial positive charge. Secondly, the rule which states that in the acid of general formula $X(OH)_mO_n$ the larger that n is, the stronger the acid, becomes readily understandable. The larger that n is, the more oxygen atoms there are in the anion $XO_{(m+n)}^{m-}$ over which the partial negative charge can be distributed, and so the less easily a proton can be attracted to any one of them. Thus sulphuric acid is stronger than sulphurous acid and nitric acid is stronger than nitrous acid.

B.2(8) DIRECTIONAL PROPERTIES OF COVALENT BONDS

By making use of yet another special assumption (hybridisation) the quantum mechanical model of molecular structure can be manipulated to generate the observed shapes of common molecules.

A simpler more elegant theory that predicts the shapes of molecules with surprising accuracy consists in arranging the electronic structure of all covalent molecules so that there should be the minimum repulsion between the electron pairs, bonding and non-bonding, that surround the central atom. Three electron pairs are directed towards the corners of an equilateral triangle (BCl_3), four pairs towards the corners of a tetrahedron (CH_4), five pairs towards the corners of a triagonal bipyramid (PCl_5) and six pairs towards the corners of a square bipyramid (SF_6). By assuming that a non-bonded pair is less localised than a bonded pair, further predictions may be made about the shapes of molecules containing both bonded and non-bonded pairs, and also about the trends of bond angles in similar compounds: NH_3, PH_3, AsH_3 for example.

B.2(9) DELOCALISATION

According to the Virial theorem (p. 81), the potential energy lowering is always the dominant factor in chemical bonding. Thus if molecular orbitals can become spread over more than two atoms, so that the electron is attracted to all the nuclei, a more stable molecule will result. This assumption (sometimes referred to as 'resonance') appears to give a plausible explanation for the extra stability of molecules like benzene, but since the distribution of any particular electron within a molecule cannot be determined experimentally, such an explanation can only remain hypothetical. The Virial theorem forecasts that the kinetic energy must rise with the fall in potential energy, but it is difficult to give any simple visual image of this— perhaps the compression of the electron into a thin flat orbital covering all the nuclei to which the electron is attracted is the least objectionable picture.

Perhaps **delocalisation** is the best term to apply to this fall in potential energy. The four circumstances in which it becomes important are as follows:
(1) When two or more atoms each possess a vacant orbital, and there is only one electron or electron pair that might fill them, the electrons will be spread over all the atoms. Examples are the bonding of metallic crystals and the dimerisation of BH_3.
(2) When one atom in a molecule has empty orbitals, and neighbouring atoms have non-bonding orbitals containing electron pairs, these pairs will be shared between the non-bonding orbitals and the vacant orbital by the formation of a new molecular orbital. Examples are the shortening of the bonds in BCl_3 and the dimerisation of $AlCl_3$.
(3) When orbitals of similar energy on neighbouring atoms overlap and form a new molecular orbital spread over several atoms. Examples are benzene, conjugated dienes, sandwich compounds and ions like NO_3^- and CO_3^{2-}.
(4) When an atom possesses a vacant orbital and there are two or more electron pairs that might fill it, it will tend to be filled partially by several electron pairs rather than wholly by one electron pair. This gives rise to very stable bonding, as in the three dimensional co-ordination polymers LiF and NaCl (see p. 97).

B.2(10) MULTIPLE BONDING: BOND ORDER GREATER THAN ONE

Multiple bonds formed by the overlap of atomic p orbitals should be *less* stable than the equivalent number of single bonds, since the closer approach of the two atoms in the former is bound to lead to an increased repulsion between the filled stable orbitals round each nucleus. For example ethylene, bond energy 611 kJ mol^{-1}, is unstable with respect to the basic polythene unit, —CH_2—CH_2—, bond energy 2 × 347 kJ mol^{-1}, and ethylene is only prevented from spontaneous polymerisation by the absence of a suitable mechanism of low activation energy. Acetylene however will polymerise readily. Graphite, each of whose carbon atoms is a member of three planar hexagonal rings, is stabilised with respect to diamond, which contains three dimensional tetrahedral bonding, because the surplus electrons in the former are delocalised throughout the whole structure.

In contrast to ethylene and acetylene, the isoelectronic oxygen and nitrogen molecules do not polymerise, a circumstance that must be connected with the low bond energy of —O—O— and —N—N— (hydrogen peroxide and hydrazine are notoriously unstable). Yet sulphur and red phosphorus, the elements in the next Period, are certainly polymerised. In this Period the non-bonding electron pairs on the atoms can be shared with the vacant $3d$ orbitals on an adjacent atom, thus giving a certain degree of double bonding to the S—S and P—P bonds: but there is no $2d$ orbital to perform the same function for the O—O and N—N bonds, so that no bonding pairs on adjacent atoms in the second Period repel each other very strongly (the very low bond energy of F—F supports this idea).

There is further evidence for the hypothesis that $3d$ orbitals participate in bonding in elements of the third Period from the considerably greater strengths of Si—O as compared to C—O, and P—O as compared to N—O. Sometimes this phenomenon is referred to as 'back-co-ordination'.

B.2(11) AVERAGE BOND ENERGY: A GUIDE TO THE FACTORS INFLUENCING THE STRENGTH OF CHEMICAL BONDS

A meaningful concept for all molecules is that of **Average Bond Energy** which at 298 K is the total heat of converting the compound to an imaginary state in which it has been broken down entirely to gaseous atoms.

For molecular compounds the average bond energy can be determined by experiment; thus for methane it is the heat of atomisation of one mole of the gas divided by 4 to give the average bond energy in kJ mol^{-1} of the single C—H bond. For infusible solids the heat of atomisation is given by the sum of the separate heats of atomisation of the component elements, less the standard heat of formation of the compound. To compare a series of compounds—the fluorides of the elements for example— it is necessary to quote the average bond energy per atom of fluorine that the molecule contains. Thus the heat of formation of CaF_2 is −1210 kJ mol^{-1}, the heat of atomisation of calcium is 180 kJ mol^{-1} and the heat of dissociation of fluorine is 155 kJ mol^{-1} so the heat of atomisation of CaF_2 is $(1210 + 180 + 155) = 1545$ kJ mol^{-1} and the average bond energy is 772 kJ mol^{-1} for the Ca—F bond.

Without any need to distinguish between the three different mechanisms of bonding, the concept of average bond energy shows that the main factors that determine the strength of any given bond are:

(1) Order
(2) Length
(3) Polarity

Thus bonding is normally stronger the greater the number of shared electrons, the shorter the bond and the more uneven the electron sharing as determined from considerations of electronegativity.

The calculation of the average bond energy of the X—F bond across Period 3 illustrates the use of this concept.

Element—F Bond	Na—F	Mg—F	Al—F	Si—F	P—F	S—F	Cl—F
Average Bond Energy (kJ mol^{-1})	753	711	682	574	489	326	251

B.2(12) DONOR AND ACCEPTOR MOLECULES; THE LEWIS THEORY

The most general definition of an acid, first proposed by Lewis, is **an electron pair acceptor** or a **proton donor**; not only does this cover H^+, which is clearly going to accept an electron pair from OH^- in the formation of water, but also molecules like $AlCl_3$ which will accept electron pairs from tertiary amines or chloride ions. The second definition covers ions like $Al(H_2O)_6{}^{3+}$, which can readily lose a proton to water (and so are acid to litmus) by forming $Al(H_2O)_5OH^{2+}$. Similarly a base is 'an electron pair donor' or a 'proton acceptor'. Clearly there is a connection here between acidity and oxidising power, for an oxidising agent is certainly an electron acceptor and a reducing agent an electron donor.

B.2(13) OXIDATION NUMBER

The concept of reduction/oxidation as gain/loss of electrons can lead to absurd conclusions, as a study of LiF shows (p. 97). It is probably better to use the entirely empirical concept of **oxidation number** for calculation purposes and reference. If hydrogen is taken as $+1$ and oxygen as -2 (except in peroxides), and all ions are made neutral by the addition of H^+ to anions and OH^- to cations, any other atom in the uncharged molecule which results can be allotted an oxidation number. Thus in the series NH_3, NH_2NH_2, NH_2OH, N_2, N_2O, NO, HNO_2, N_2O_4, $NO_3{}^-$, the oxidation state of nitrogen varies by steps of one unit from -3 to $+5$.

B.3
Electronic justification for the modern shape of the Periodic Table

B.3(1) SEPARATION OF MAIN GROUP ELEMENTS FROM TRANSITIONAL GROUPS

The d orbitals of Main Group elements are always empty or completely filled.

The compounds of the highest oxidation state of the Transitional Group elements up to T7 (manganese) contain no electrons in atomic d orbitals; thus perbromates have the same type of molecular orbital as permanganates, the difference being that the former has its $3d$ atomic orbitals completely filled with electrons.

Zinc differs from calcium in having a set of filled $3d$ orbitals, electrons from which are never involved in chemical bonding. The additional nuclear charge associated with these ten electrons holds the two penetrating $4s$ electrons of zinc far more strongly than those of calcium; thus the zinc ion is smaller and it is more difficult to ionise the $4s$ electrons. This results in a lower place in the electrochemical series and a greater polarising power for the zinc ion—suppositions which are very accurately borne out in the chemical differences between zinc and calcium.

B.3(2) THE CONTINUITY OF THE MODERN TRANSITIONAL SERIES

Each of the elements from T3 to T11 differs from the preceding one by the addition of a single d electron, which leads to a comparatively regular gradation in all chemical properties. There is no justification for any separation of Fe, Co, Ni from the remainder.

B.3(3) INNER TRANSITIONAL ELEMENTS

The extremely small changes in properties between one inner transitional element and the next correspond to the successive addition of one electron to the f orbitals deeply buried within the atom, whose interaction with the chemical environment is minimal. Very strong oxidising agents will remove the single f electron from cerium. All the elements between La and Lu will contain unpaired electrons and show paramagnetism (see Hund's Rule, p. 88), and the $4f^7$ configuration of Gd(III) justifies its predicted stability.

B.3(4) DISCONTINUITY IN THE LATER MAIN GROUPS

The effect of the filled $3d$ orbitals on the later Main Group elements of Period 4 as compared to Period 3 will be to bind the peripheral electrons more strongly owing to the increased nuclear charge working upon the penetrating parts of their orbitals. This will raise their electronegativities, so that there will be a reluctance to form compounds containing the element in its highest oxidation state.

B.3(5) THE UNIQUE POSITION OF HYDROGEN

The anomalous properties of hydrogen depend upon the fact that it can be classified in electronic terms in three different ways.

(a) Containing one electron in excess of a stable structure (like an alkali metal).
(b) Lacking only one electron to reach a stable structure (like a halogen).
(c) Having its available low energy orbitals half-filled (like an M4 element).

It cannot therefore be fitted satisfactorily into any one Group of the Periodic Table. It is best placed above and a little to the left of carbon, since it is slightly less electronegative than the latter.

B.4
General structure of the whole Periodic Table: trends among the Groups

B.4(1) THE ELEMENTS

(a) Metals are formed when the number of electrons available for bonding is less than the number of low energy orbital vacancies, with consequent delocalisation of the electrons throughout the whole metallic crystal (p. 102). The more electrons there are, the more strongly the metal ions are bonded together, with consequent increases in density, hardness and melting point across the Periods.

Boron, being the smallest element in Main Groups 1 to 3, has a unique structure (p. 125) which allows extensive delocalisation, but does not permit the entirely free movement of electrons which is required for electrical conductivity.

(b) The more unfilled orbitals of low energy there are available (which for transitional elements involves d orbitals also) the more possibilities there are for delocalisation. Thus the transitional and inner transitional elements are all metals.

(c) The metallic structure of the later Main Group elements can be accounted for in two ways. First, the reluctance of the $5s$ and $6s$ electrons to ionise limits the number of delocalised electrons for which orbital vacancies must be found. Second, elements of high atomic number have nd orbitals available of very little greater energy than the ns and np orbitals, which can accept delocalised electrons.

(d) Each carbon and silicon atom has four strong directional bonds to four different neighbours in the lattice of the element. The disruption of this stable three dimensional array requires very high energies.

(e) The volatile non-metals all involve directional covalent bonds and form discrete molecules in the vapour state:
e.g. P_4, As_4; S_8, Se_8; F_2, Cl_2, Br_2, I_2. The four phosphorus and arsenic atoms are at the corners of a regular tetrahedron, and the eight sulphur and selenium atoms form puckered rings. The smaller

the atoms, the stronger the bonds that are formed to hold the molecules of the elements together:
H—H, 431 kJ mol⁻¹; C—C, 347 kJ mol⁻¹; Cl—Cl, 242 kJ mol⁻¹
I—I, 151 kJ mol⁻¹. [F₂ is exceptional (p. 139)].

The boiling points of these elements depend upon the ease with which one molecule with temporarily displaced electrons can polarise its neighbours. These *van der Waals forces* increase down the Group owing to the increasing polarisability of large molecules.

Nitrogen and oxygen overlap their atomic p orbitals to form double bonds which are more than twice as strong as —N—N— and —O—O—, the weakness of which is accounted for by the repulsion of the non bonding electron pairs (p. 134).
(f) Elements which dissolve in both acids and alkalis do so by virtue of the amphoteric character of their oxides.

B.4(2) HYDRIDES OF THE ELEMENTS

The concept of electronegativity provides a very clear interpretation of the Periodic trends of the compounds of hydrogen with the elements. Since hydrogen has but one low energy orbital it cannot form a covalent bond involving a pair of electrons from another atom without producing a charged ion; thus $NH_3 + H^+ \rightarrow NH_4^+$; H_4O^{2+} and H_4F^{3+} are too unstable to exist.

Crossing the Period from lithium to fluorine involves a transition from a salt-like solid in which hydrogen carries a considerable partial negative charge, through almost purely covalently bonded hydrogen compounds with no charge to acidic compounds in which the hydrogen carries a partial positive charge.

(a) Lithium-hydrogen bonds are polar, so that association between separate molecules of LiH takes place. The valence electrons on the hydrogen are delocalised through the available orbitals on the lithium, leading to a close-packed structure in which the lithium nuclei are far closer together than they are in lithium metal.

(b) The beryllium atom is less electropositive than lithium, so the linear BeH_2 does not form salt-like crystals which contain free ions when molten, but the two unoccupied orbitals on the beryllium atom can become filled by the two electrons already involved in a Be—H bond, so that these two electrons are attracted to three nuclei. This 'hydride bridging' can only take place when the hydrogen carries a partial negative charge: it is really a limited form of the extensive delocalisation in LiH and also leads to three dimensional structures which are consequently involatile. BH_3 has only one free orbital, and boron is more electronegative than beryllium; yet the hydrogen is still sufficiently negatively charged to act as a hydride bridge, two BH_3 groups combining to form B_2H_6. In all these compounds the partially negatively charged hydrogen will react with H_3O^+ in water to form hydrogen gas.

$$LiH + H_3O^+ \rightarrow Li^+ + H_2 + H_2O$$

(c) Transitional metal hydrides are interstitial compounds, high melting, stable and non-stoichiometric; the small atoms of hydrogen pack into the holes in the metallic crystal lattice. Palladium hydride, $PdH_{0.6}$, is used as a membrane for purifying hydrogen gas.

(d) Carbon is slightly more electronegative than hydrogen so the hydrogen in the tetrahedral CH_4 has a small partial positive charge. This means that it is resistant to oxidation (since it will not readily lose electrons) but it is not sufficiently positive to show acidic properties. The resulting polarity of the symmetrical molecule CH_4 is zero, so there is little intermolecular attraction and methane is a very low-boiling gas. In the pyramidal ammonia molecule, nitrogen is more electronegative than hydrogen, so there is considerable intermolecular attraction and in addition some hydrogen bonding (protonic bridging, p. 135) in which the small partially positively charged hydrogen atom is attracted to two partially negatively charged nitrogen atoms. The unshared pair of electrons on the nitrogen atom is sufficient to make ammonia predominately basic, yet the partial positive charge on the hydrogen means that it can show some acidic properties.
Thus:

$$NH_3 + H^+ \rightarrow NH_4^+ \text{ but } Na + NH_3 \rightarrow Na^+NH_2^- + \tfrac{1}{2}H_2$$

Oxygen is more electronegative than nitrogen so the V-shaped water molecule is more acidic than ammonia; hydrogen bonds are even more easily formed and water boils at a higher temperature. Finally, in the compound of hydrogen with fluorine, the most electronegative element, hydrogen carries a large partial positive charge so that anhydrous HF is a strong acid and H_2F^+ exists only to a minute extent in the liquid. There is very strong hydrogen bonding, but since HF contains only one hydrogen atom the associated molecules so formed can only be linear or cyclic, not three dimensional; this accounts for the boiling point of HF being lower than that of H_2O.

Like most other covalent compounds (p. 104), the strength of bonds to hydrogen increases with the electronegativity difference of the two elements but decreases as the two atoms become of increasingly different size. The combination of these two factors reinforce on another for the hydrogen halides, as the following bond energies (kJ mol^{-1}) show: H–F, 560; H–Cl, 431; H–Br, 364; H–I, 296.

B.4(3) OXIDES OF THE ELEMENTS

(a) In the early Main Groups metal-oxygen bonds are very highly polar, and therefore will tend to associate together, just as the salt-like hydrides do, and form three dimensional co-ordination polymers in which all atoms of each kind have the same environment; for example in calcium oxide each calcium atom is surrounded by six oxygen atoms, and each oxygen is surrounded by six calcium atoms so that no individual molecules can be detected. In oxides of elements of intermediate electronegativity, the molecule is still strongly polar, and association takes place: ZnO is an example of such an oxide in which the environments of the atoms it contains are not symmetrical: there are distinct groups of molecules within the crystal. In M4, SiO_2 contains directional covalent bonds arranged in a diamond type lattice, each C—C bond being replaced by an Si—O—Si bond.

As the electronegativity of the element approaches that of oxygen, the tendency to form discrete covalent molecules grows. P_4O_{10} has four phosphorus atoms at the corners of a tetrahedron with six oxgyen atoms on each edge and four oxygens projecting from the apices. SO_3 is a planar Y-shaped molecule, H_2O and Cl_2O are V-shaped.

Discrete molecules with no strong intermolecular bonds in the solid are much more easily melted and volatilised than three dimensional covalent lattices or three dimensional co-ordination polymers; there is in fact no real distinction between the latter two structures except the polarity of the bonds they contain.

Unlike hydrogen with only one electron vacancy, oxygen with two vacancies can accommodate electron pairs from donor atoms to yield oxides; this process can be repeated until the element is in its highest oxidation state. The conversion of SO_2 to SO_3 is a good example. In these oxides the electronegativity difference between oxygen and the element involved will be shared out among all the oxygen atoms, so the net effect will be to lower the partial negative charge on each of them. Now oxygen, being electronegative, will normally seek to increase its partial negative charge; thus the lower the partial negative charge on oxygen in a particular compound, the more likely that compound is to be an oxidising agent. Thus the oxides of the early Main Group elements, in which oxygen has a high partial negative charge, have no oxidising properties: whereas the oxides of the Group oxidation state of the later Main Group elements are all oxidising agents, Cl_2O_7 for example. Some of the lower oxides of the later Main Group elements are also oxidising agents for the same reason, e.g. I_2O_5. In the extreme case of F_2O the partial charge on the oxygen is positive, and this compound is very unstable and an extremely strong oxidising agent.

(b) The generalised form of the Lewis Theory of Acids and Bases (p. 104) makes it clear how close is the connection between acidity and oxidising power, basicity and reducing power. The combined oxygen in an oxide of an early Main Group metal carries a large partial negative charge, and therefore it is likely to act as an electron donor (equals proton acceptor equals base). Similarly the central atom of a non-metallic acidic oxide, which has its own electrons partially removed by the surrounding electronegative oxygen atoms, will be an electron acceptor (equals proton donor equals acid)— particularly if it has low energy orbital vacancies available. Thus oxides of the early Main Groups, in which the oxygen atom is a potential electron donor, will tend to react spontaneously with oxides in the Main Groups whose central atom is a potential electron acceptor, and salts will be formed. Thus,

$$CaO + SO_3 \rightarrow CaSO_4$$

(c) Water, a typical amphoteric oxide, will act as an electron acceptor for early Main Group oxides and an electron donor to later Main Group oxides. The compounds formed will be ionic, and will dissolve in excess water owing to its high dielectric constant. Neither of these reactions will take place to any great extent with oxides of elements of intermediate electronegativity, which therefore do not form ionic compounds and are consequently insoluble.

The separation of hydroxide ions from hydroxides of metals in M1 and M2 requires that an electron originally shared with the rest of the molecule must become exclusively the property of the hydroxide ion. This can only happen if the bond between the oxygen and the central atom is already quite polar, with considerable partial negative charge on the oxygen.

The empirical rule [A.5(3)] governing the strength of acids is easy to interpret in terms of the concept of electronegativity. The more electronegative oxygen atoms there are in competition for the electrons of the single central atom, the less successful each one of them will be in acquiring partial negative charge. Since protons will not separate from an oxygen atom with a high partial negative charge, the more oxygen atoms and the fewer hydroxyl groups there are in an oxyacid the stronger that acid will be. Thus HNO_3 is a stronger acid than HNO_2, H_2SO_4 is a stronger acid than H_2SO_3, and $HClO_4$ is stronger than the other oxyacids of chlorine. The rule also applies to ions: H_3PO_4 is stronger than $H_2PO_4^-$ which is stronger than HPO_4^{2-} because the overall charge on the whole ion increases the partial negative charge on the oxygen. Similarly in the isoelectric series H_4SiO_4, H_3PO_4, H_2SO_4, $HClO_4$, as the central atom becomes more electronegative the partial negative charge on the peripheral oxygen atoms will fall and the acids will become stronger.

(d) Oxides of intermediate polarity will dissolve in solutions having either an excess of H_3O^+, or an excess of OH^-; those which dissolve in both these solutions (but scarcely at all in water) are truly amphoteric. It is at first difficult to see why not all oxides of elements of intermediate electronegativity are amphoteric; the answer probably lies in the very wide range of structures which hydroxides can adopt. The relative sizes of the metal atoms and the degree of polymerisation make each hydroxide system unique, and may prevent it reacting by one or more of the mechanisms which control acidity or basicity. The hydroxides which are unambiguously amphoteric are those which have alternative co-ordination numbers of four or six; e.g. $Al(H_2O)_6^{3+}$ and $Al(OH)_4^-$. Emphatically amphoteric character does not depend on one single property like electronegativity or relative electrode potential.

Oxides of elements of intermediate electronegativity which dissolve only in acid tend to be in the early Main Groups, like MgO, whereas those which dissolve only in alkalis tend to be in the later Main Groups, like SiO_2.

(e) In CrO the partial negative charge on the oxygen is sufficiently high to make it a basic insoluble oxide which invites the attack of protons, whereas in CrO_3 the share of partial negative charge on each oxygen is low enough to make this substance strongly acidic.

(f) The more a complex oxyanion can control its electrons, the more stable it will be. The proton in anhydrous acids has a very high charge/radius ratio, and so is highly polarising; this accounts for the instability of $HClO_4$ as compared with $KClO_4$. For the same reason the salts of hydrated cations are more stable than those of the same cation when anhydrous, which is smaller but has the same charge.

B.4(4) HALIDES OF THE ELEMENTS

(a) In compounds where the partial negative charge on a halogen is high it will attempt to share its non-bonding electron pairs with the empty orbitals of other atoms (normally electropositive metals) with which it is combined. Thus six lithium ions surround each halogen ion, and six halogen ions surround each lithium atom in the lithium halides. As the electronegativity of the element with which the halogen is combined falls, the bond becomes less polar and weaker, so the molecule becomes more volatile and unstable.

(b) The polarising power of the Al^{3+} ion is very similar to that of the transitional metal ions: these metals have chlorides, bromides and iodides which readily sublime, and react exothermically with water to form the hydrated ion of the metals, which then readily expels a proton.

(c) The lower the partial negative charge on the halogen atom, in a particular halide, the more likely it is to seek electrons elsewhere by forming compounds with other more electropositive atoms. The fluorine atoms in SbF_5 will carry a smaller partial negative charge than the fluorine atoms in SbF_3, since in the former there are more fluorine atoms competing for the electrons of the central antimony atom: thus SbF_5 will be the better fluorinating agent.

(d) For halides of a particular electropositive metal, the fluoride will be the most polar owing to the higher electronegativity of the fluorine atom, so the lattice energy will be higher and the fluorides will tend to be less volatile and less soluble than the other halide salts.

Among the volatile halides of non-metals, the iodide is the largest molecule and the most polarisable, and hence forms the strongest van der Waals bonds. Thus it has the highest boiling point.

As the electropositivity of the metal falls, crystals are formed in which not all the atoms have a symmetrical environment—a change which foreshadows the formation of discrete covalent molecules in the crystal. Fluorine in the fluorides carries a higher partial negative charge than the other halogens, so in comparison with them it will be more reluctant to give up the polymeric nature of structures like LiF, in which each fluorine atom can be considered as a bridge between three pairs of lithium atoms. Elements of intermediate electronegativity retain this bridging in their fluorides, with its consequent three dimensional structure and low volatility, long after the other halides have become covalent volatile liquids. Such structures do not permit the free rotation of ions in the heated solid, so they melt at high temperatures; but once the fluoride bridges are broken, the melt tends to contain covalent molecules rather than free ions—which volatilise immediately. Thus Na_3AlF_6 has a long molten range and the liquid contains free ions; whereas AlF_3, in which the octahedral AlF_6 groups are linked by fluorine bridges, sublimes at 1270°C.

B.5
Theoretical interpretation of trends within Periodic Groups

B.5(1) GROUP M1: THE ALKALI METALS

The straightforward nature of M1, lithium and the alkali metals, allow the effects of increasing size and mass on physical and chemical properties to stand out very clearly. First, atomic size increases down the Group, and so therefore it is easier to remove an electron; thus the elements react increasingly readily. The binding energies in the close packed metal lattices are relatively weak, because there is only one valence electron per metal atom; the metals are consequently increasingly soft, increasingly low melting and increasingly good conductors of electricity. The first ionisation energy of lithium is higher than that of the remaining elements, so its position in the electrochemical series should be lower. However, the effect of the very high heat of hydration of the lithium ion creates an anomaly (p. 94), and lithium is in fact the most electropositive of all the elements.

Second, and far more important chemically, is the decrease in polarising power of ions down the Group. An approximate quantitative estimate of polarising power may be obtained from the ratio charge/radius for a particular ion (see p. 99). Polarising power is thus greatest for the small lithium ion, and the onset of covalent character is important in the following ways.

(a) The more polarising the cation, the more strongly will it attract the non-bonding electron pairs on the oxygen of a water molecule and the more strongly hydrated will be the ion. If both cations and anions are small, the lattice energy of the crystal will be high; although the heat of hydration is also high for small ions it will probably not be sufficient to disrupt the lattice (LiF is insoluble).

The lattice energy of the salts of large cations and very large anions is not very great, but the energy of hydration of the large cation is so small that it is insufficient to disrupt this weak lattice. Thus the perchlorates (ClO_4^-) and hexanitrocobaltate(III) ($Co(NO_2)_6^{3-}$) compounds of K, Rb and Cs are all insoluble.

(b) The more polarising the cation, the more it will deform a complex oxyanion of high charge like CO_3^{2-}, tending to form the more compact metal oxide and CO_2. This is especially true of weak acids, the oxygen atom of which is already carrying a moderate negative charge (see p. 101); otherwise it would release a proton more easily! Thus the salts of weak acids tend to be thermally unstable and decompose before they melt if the cation is sufficiently polarising. Similarly they tend to have low solubilities owing to increased bond strength; e.g. Li_2CO_3 is insoluble.

Only the least polarising cations can form salts with the most polarisable anions, particularly O_2^{2-}, O_2^- and I_3^- for such anions are easily decomposed by strongly deforming cations.

(c) The heats of formation of the fluorides, oxides and hydrides are greater for lithium because the close approach of these small ions in the crystal leads to a high lattice energy. However owing to the lower electronegativity of lithium, its bonds to the halogens are not so polar as those of the other M1 metals, so it comes as no surprise that the lithium halides are the most volatile in the Group.

(d) The non-polar nature of the lithium alkyls, and the comparatively large heat of dissociation of the bond in Li_2 clearly show the onset of covalency.

SOLUBILITY OF SALTS IN GROUP M1

The tendency for a salt to dissolve depends upon two factors; the heat of solution and the degree of order which is lost when a highly organised crystal dissolves. The solubility will be greater the more heat that is given out when the salt dissolves, and the greater the net decrease in order. Both the heat of solution and the ordering factor consist of two opposing parts. The heat term is given by the lattice energy of the crystal minus the heat of hydration of the ions—two large factors of approximately the same size; the ordering effect is made up of the inevitable loss of order when a highly organised crystal is broken down by solvent molecules, counterbalanced by the gain in order stemming from the loss of freedom of those water molecules incorporated into the hydration sheath round the freed ions.

Consider the ordering effect first. The loss of order when a crystal breaks up is much the same for all compounds; the gain in order associated with a hydration sheath is larger for ions for which the value of the ratio charge/radius is high. Thus from ordering considerations alone large ions of low charge should be soluble. This is generally true; all nitrates and the vast majority of the salts of M1 metals are indeed soluble. Hydrated salts are always soluble, for the ordering effect on the dissolved ions has already taken place in the crystal, thus dissolution is bound to involve a large decrease in order.

The higher the lattice energy of particular crystal, the less likely is that crystal to dissolve in water. The lattice energy for cations and anions which are small and of similar size is rather high, but decreases according to Coulomb's Law as both ions become larger. Thus salts containing ions of similar size are often more insoluble than other members of the Group, and the more dissimilar are the ions in size the greater the lowering of the lattice energy and the more likely the salt is to be soluble: e.g. LiF and $KClO_4$ are insoluble, while KF and $LiClO_4$ are soluble. The heat of hydration decreases as the value of charge/radius grows smaller; and a low heat of hydration does not favour solution, especially if the ions are of similar size.

The salts of weak acids normally contain rather large oxyanions with a high partial negative charge on the oxygen. Combined with cations which are polarising, the incipient covalency leads to a high lattice energy and consequent insolubility (RbCl soluble, AgCl insoluble). For larger cations which are less polarising the better matching in size increases the lattice energy, but the polarisation is decisive; thus Li_2CO_3 and Li_3PO_4 are insoluble, but the other M1 carbonates and phosphates are soluble.

The hydration of salts of strong acids depends upon the polarising power of the cation. The hydration of salts of weak acids depends upon whether the cation, by becoming hydrated, can grow to be as large as the anion, thus increasing the lattice energy of the crystal and making it more stable. In M1 the hydration factor dominates, but in M2 the lattice energy factor is more important, especially for salts of weak acids.

B.5(2a) GROUP M2: THE ALKALINE EARTHS

As in M1, the main trends in the elements Mg, Ca, Sr, Ba depend firstly upon the increase in size and consequent decrease in ionisation energy down the Group, and secondly on the decreasing polarising power of the larger ions. However with divalent ions polarisation is bound to be greater than in M1, which makes predictions of lattice energy in accordance with Coulomb's Law much more hazardous.

The decreasing ionisation energy accounts for the increasingly soft nature of the elements; however, they are not so soft or low melting as the M1 elements, for they contain two electrons per atom to bind the metallic lattice together. They are sufficiently electropositive to form ionic compounds with nearly all the non-metals, which hydrolyse in solution (the stability of the nitride of magnesium, as well as that of lithium and aluminium is probably connected with the small size of the N^{3-} ion).

The greater thermal instability of the salts of weak acids, the covalent nature of the magnesium alkyls, the strong hydration of Mg^{2+}, the multiplicity of magnesium complexes and the comparative instability of the peroxides can all be explained in terms of a higher charge/radius ratio than the corresponding elements in M1.

The beryllium cation—if it existed—would be exceedingly small, and the chemistry of beryllium shows the marked covalent character expected for an ion whose charge/radius ratio is very high.

SOLUBILITY OF SALTS IN GROUP M2

The M2 metal sulphates are matched for charge, so the lattice energy is high and the match for size improves down the Group. Thus the lattice energy of these compounds increases as the heat of hydration decreases, so the sulphates are increasingly insoluble.

The salts of weak acids are generally insoluble owing to the polarisation effect; the larger the value of charge/radius for the cation, the more insoluble they should become. However, the matching of cation/anion sizes is far more important in M2 salts than in M1 salts; since the small magnesium cation is far smaller than the larger anions, magnesium salts are sometimes less insoluble than those of calcium—the oxalate for example. However, with the smaller ions of weak acids, like O^{2-} and OH^-, the compounds with magnesium are the most insoluble in the Group.

As in M1, the hydration of the salts of strong acids depends largely upon charge/radius for the cation. However, whether the salts of weak acids contain water of crystallisation is dependent upon the existence of cations and anions that are matched for size and charge. The crystallisation of a hydrated salt from solution is evidence that the lattice energy of the hydrated salt is higher than that of the anhydrous salt.

B.5(2b) METALS IN THE LATER MAIN GROUPS

The ions showing the Group oxidation number in M2′ and M3 have a pseudo-noble gas structure; Zn^{2+} for example is /argon core/$3d^{10}$/$^{2+}$. The energy required to remove electrons from the filled d orbitals is not prohibitively high, but it is too great to be surmounted by any chemical reaction yet known when the positive ion has a double charge. Again, the configuration of the Pb^{2+} ion is /xenon core/$5d^{10}$/$6s^2$/$^{2+}$, similar to that of the mercury atom. The penetrating nature of the $6s^2$ electrons, and the large increase in the positive charge on the nucleus corresponding to the lanthanides means that the energy required to remove the $6s^2$ electrons from Pb^{2+} and so form Pb^{4+} is far too great to be brought about by a purely chemical process. Cations with four positive charges are rare enough at any time, and for lead their formation is impossible.

Thus the Group oxidation state of all elements from M4 to M7 will always involve covalent bonding; it is instructive to consider what factors operate to stabilise the Group oxidation state with respect to any lower state, AsF_5 as opposed to AsF_3 for example. The electronic structure of arsenic is /argon core/$4s^2$/$3d^{10}$/$4p^3$; the less penetrating $4p$ electrons are held less strongly than the $4s$ electrons which spend much of their time near the nucleus [by this time the closed $3d$ rhythmic pattern is very stable (see p. 143)]. In order that any bonding shall take place, the three $4p$ electrons must overlap with the atomic orbitals of fluorine to form compound molecular orbitals of lowered energy. To form five bonds requires the additional participation of the $4s$ electrons: they must be **promoted** into a suitable orbital (a process which requires energy) before they can enter a bonding molecular orbital with consequent energy lowering. If the energy required for promotion is less than the energy recovered from bonding, the overall process will be favourable and AsF_5 will be formed from AsF_3 and fluorine.

High temperatures always favour disorder, so that the formation of the compound containing the greater number of atoms becomes progressively more unlikely as the temperature rises. Thus incipient oxidation states of $+1$ and $+2$ among the earlier elements in M3 and M4 will make its appearance first in transient compounds formed at high temperatures.

The reason why the elements of high atomic weight in the later Main Groups are more inclined to show oxidation states lower than that of the Group number is straightforward. The strength of covalent bonds decreases (provided the polarity of the bond does not change much) as the bonds get longer, which they are bound to do in the analogous compounds in the same Group from Period 3 to Period 6, so the energy available for the promotion of ns electrons steadily falls down the Groups. In Period 6 the extra reluctance of the $6s^2$ electrons to ionise actually raises the energy necessary for promotion, making the Group covalency even more unlikely.

The formation of covalent compounds of the lower oxidation states of the later Main Group elements leaves empty orbitals available, so the production of complex ions like $PbCl_4^{2-}$ and BiI_4^- is not surprising.

Cations of charge greater than $+3$ are not formed, so that there is no competition between covalent and ionic bonding in the Group oxidation states of M4 and M5. However, with the lower oxidation states the energy difference between volatile molecular compounds with fairly polar covalent bonds, and a distorted three dimensional ionic lattice may be small: for $Pb(II)$ and $Bi(III)$ the ionic lattice is favoured for anions like F^- and SO_4^{2-} which are unwilling to donate electron pairs to the metal atom. Indeed, when molten, PbF_2 will conduct electricity, which shows that the Pb^{2+} ion can actually exist free.

The later Main Group metals form oxides and sulphides which are more stable than can be accounted for by simple electrostatic attractions in an ionic crystal. In addition they are mostly brightly coloured (a phenomenon usually associated with charge transfer) and some of the sulphides have a metallic sheen. This suggests that the extra strength of the bonds that hold them together must come from a novel non-ionic mechanism not available to the M1 and M2 metals. A possible hypothesis is that the non-bonding electron pairs on the oxide or sulphide ion can form bonding molecular orbitals with the p orbitals on the metal atoms with consequent partial delocalisation of electrons throughout the lattice (the p orbitals for the later Main Group elements are stabilised by the extra protons corresponding to the filled d shell). Such a mechanism is not unlike the 'back-co-ordination' which is used to account for the high strength of the Si—O and P—O bonds (p. 131).

The non-bonding electron pairs on the sulphide ion are less strongly held than those on the oxide ion since the former is larger, so it is not surprising that the heat of formation of the sulphides is nearly as high as that of the oxides, a feature that occurs nowhere else in the

Periodic Table. This means that the sulphides will be insoluble in acids (p. 136) and will be formed in nature in preference to the oxides. Besides the stability of sulphur complexes like SCN^- and $S_2O_3{}^{2-}$, the insolubility of iodides like PbI_2 and the stability of iodine anionic complexes like $HgI_4{}^{2-}$ also support the hypothesis of electron pairs donated by the anions, since iodide will be much more likely to do this than the almost non-polarisable fluoride ion. Nitrogen is not very polarisable either, which explains the lack of eagerness to form ammines: in cases where they might be found the acidity of the solution of the extensively hydrolysed salt might well be sufficient to decompose them in the following way:

$$Zn(NH_3)_4{}^{2+} + 4H_3O^+ \rightarrow Zn(H_2O)_4{}^{2+} + 4NH_4{}^+$$

It is difficult to account for the extensive formation of basic salts by the B metal cations without invoking the same hypothesis. The heats of hydration of Pb^{2+} is very little different from those of the M2 metals, and lead is often reluctant to incorporate water of crystallisation into its salts—witness the anhydrous nitrate used for the laboratory preparation of N_2O_4. Yet lead very frequently forms salts such as $PbO.PbCO_3.H_2O$, tin is extensively hydrolysed in solution according to the equation,

$$Sn(H_2O)^{2+} \rightarrow SnOH^+ + H^+,$$

and $BiCl_3$ solutions hydrolyse on dilution giving $BiOCl$. These results become comprehensible in terms of the available electron pairs on the oxide anion, and the strong bond thus formed with the later Main Group elements.

This ability to accept electrons from polarisable anions is shared by elements late in the transitional Periods which also have high electron affinity. Thus copper has an insoluble sulphide and silver thiosulphate complexes decompose to give Ag_2S. Unlike the Transitional elements, however, the later Main Group metals will not form complexes with unsaturated ligands, for they are unwilling to donate electron pairs from their increasingly tightly held d orbitals into molecular orbitals formed from empty orbitals on the ligands.

B.5(2c) GROUP M2′

The M2′ metals are smaller and more dense than their M2 analogues owing to the increased nuclear charge accompanying the ten extra $3d$ electrons they contain. In zinc, the penetrating $4s$ electrons will be attracted much more strongly by the nucleus than they are in calcium, so the electron cloud will be less diffuse. For the same reason the ionisation energy for the $4s^2$ electrons in zinc is considerably higher, so that the electrons will be less delocalised in the metal lattice and the metallic bonding will not be so strong. Despite the fact that the smaller zinc ions will be more polarising than Ca^{2+}, this effect is not sufficient to offset the large rise in the ionisation energy, so the electrode potential of zinc is much lower than that of calcium.

This effect is even more striking with the $6s^2$ electrons in mercury, which have no less than 24 extra protons to attract them compared to the equivalent M2 metal barium (see p. 118). The ionisation energy is very high (mercury exists as a monatomic vapour), so mercury has very weak metallic bonding, is very low in the ECS and is of higher electronegativity than cadmium. The oxidation state of mercury(I) compounds is accounted for the diatomic Hg_2^{2+} ion, which contains one atom of the metal in oxidation state $+2$ and one atom of the metal in oxidation state zero. This unwillingness of the $6s^2$ electrons to ionise in the later Main Group elements is often called the **'inert pair' effect,** but it is not wholly understood.

For zinc and cadmium, the covalent nature of the alkyls, the covalent structure and Lewis Acid properties of the anhydrous zinc halides, the strong hydration of M^{2+} and the hydrolysis of the salts of strong acids, the thermal instability and basic nature of the rather insoluble salts of weak acids, the great number of complexes and the decreasingly amphoteric nature of the hydroxides can all be confidently ascribed to a comparatively high charge/radius ratio for the M^{2+} ion (for a comparison of Zn and Ca, see p. 57).

The chemistry of mercury is dominated by the instability of the bond between mercury and oxygen, for which there is no adequate explanation. HgO itself has a very low heat of formation and is thermally unstable; it is generated by any reaction which might be expected to produce $Hg(OH)_2$ and there are no signs of any acidic properties. It is rather insoluble and is also a very weak base—perhaps there is no mechanism of low activation energy by which it can be attacked. The salts of the strong acids are hydrated, as would be expected for a polarising cation: but these ions very readily form complexes in which the water molecules are displaced—oxygen will practically never act as a donor to mercury. The low heat of formation of the bond between mercury and oxygen accounts for the anomalous stability of the mercury alkyls to air and water.

The structure of the ion of the lower oxidation state of mercury has been established to be $[Hg(o) - Hg(\text{II})]^{2+}$ in the following ways:
(a) By the recognition of a stretching frequency for the Hg—Hg bond in the Raman spectrum of solutions of mercury(I) nitrate.
(b) For the reaction:

$$Hg° + Hg^{2+} \underset{\longrightarrow}{\overset{\longleftarrow}{}} 2Hg^+$$

the equilibrium constant is $K = [Hg^+]^2/[Hg^{2+}]$.
However, for the reaction,

$$Hg° + Hg^{2+} \rightleftharpoons Hg_2^{2+}$$

the equilibrium constant is $K' = [Hg_2^{2+}]/[Hg^{2+}]$.

For different concentrations of mercury(I) and mercury(II) ions in presence of the free metal, it is the latter relationship that is found to be constant.

In the reaction:

$$Hg_2^{2+} \rightleftharpoons Hg^{2+} + Hg°$$

anything which can reduce the concentration of Hg^{2+} in solution, particularly complex formation or precipitation, will favour a change towards the right hand side of the equation by Le Chatelier's Principle and so disproportionation will be the result. Hg^{2+} is more polarising than Hg_2^{2+} so it will form complexes more easily; thus any addition of ligands like CN^-, I^-, SCN^- or NH_3 to the equilibrium solution will lead to the preferential formation of mercuric complexes and the disproportionation of any Hg_2^{2+} ions that may be present. If chloride ions are added to the equilibrium solution, the very insoluble Hg_2Cl_2 is precipitated in preference to the slightly ionised $HgCl_2$. Similarly the stability of mercury(I) nitrate solutions is dependent upon the fact that mercury(II) nitrate is also extensively ionised.

B.5(3a) ALUMINIUM AND THE ELEMENTS OF M3

The ionisation energies of the M3 metals are all rather similar, so no very well defined trends would be expected, and the electronegativities are irregular, anomalously high for Ga and Tl marking the enlargement of the rhythmic pattern from 8 to 18 and from 18 to 32 orbital vacancies. As the ions get larger (though the increase is not very great owing to the extra protons associated with the transitional and inner transitional elements) their polarising power gets less. Thus the heat of hydration, very high for aluminium, will diminish down the Group, and this will lower the standard electrode potentials of the other elements considerably (p. 93). The first ionisation energy, corresponding to the first p electron, is not high, so there are slight tendencies towards an oxidation state of $+1$ in the elements before thallium, where the 'inert pair' effect makes $+1$ the stable oxidation state. The ionisation energy of the boron atom, even with its small size and consequent large lattice energy, is far too high to allow the formation of B^{3+}.

The high melting point of the metallic fluorides depends upon the comparatively small size of the ions, rather than upon their extensive ionic character. The other halides of the M3 metals are all covalent, and they are strong Lewis acids—hence their catalytic properties. The empty orbital on the metal atom can be filled by one of the non-bonding electron pairs on a chlorine atom.

The structure of $Al_2(CH_3)_6$ is probably similar to that of aluminium chloride, the two electrons binding the electronegative methide group to the electropositive metal being shared with both aluminium atoms making the compound electron deficient. Beryllium behaves in a similar way.

The polarising power of the small Al^{3+} is very great, and aluminium salts are thus invariably hydrated. The small size of both atoms makes the Al—O bond, like the Be—O bond, very stable indeed, which accounts for the formation of basic salts, the amphoteric character of $Al(OH)_3$ and the expulsion of protons from $Al(H_2O)_6^{3+}$ in solution.

Octahedral complex formation is not possible for elements of the second period, but elements of the third and later Periods can form molecular orbitals involving 12 valence electrons.

B.5(3b) BORON

The extreme stability of elemental boron depends upon the structure. This consists of interlinked B_{12} units, each of the twelve boron atoms being at the apices of an icosahedron. Since each atom takes part in boron-boron bonding both within the icosahedron and outside it, every atom of boron participates in far more bonds than it has bonding electrons, which naturally involves extensive delocalisation, without, however, allowing the entirely free movement of electrons which would lead to electrical conductivity.

The structures of the electron-deficient boron hydrides become readily understandable once the boron atoms within them are visualised as fragments of the original icosahedral B_{12} unit. B_2H_6, the simplest of them, contains two bonds in which a pair of electrons is delocalised over three atoms, while B_4H_{10} contains four of these bridging hydrogen atoms.

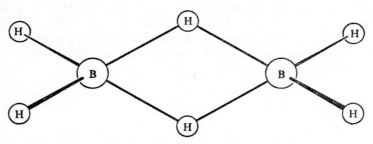

The bridging hydrogens are entirely different from the others, for which alkyl groups can be substituted. The $B_{12}H_{12}^{2-}$ ion, predicted five years before it was discovered, is rather stable because of its icosahedral structure: the BH_4^- ion is isoelectronic with methane, and is also tetrahedral.

BF_3 is the strongest known Lewis acid (p. 104), since there is an empty orbital of very low energy which readily accepts an electron pair on the boron atom. At first sight it is surprising that the boron halides are not dimeric, but measurement of the B—Hal bond lengths indicates that they are far shorter than the value calculated for single bonds. This is interpreted to mean that the electron pairs on the halogen atoms which are normally non-bonding alternately fill the empty orbital on the boron; thus, the B—Hal bonds have partial double bond character (p. 102).

The pseudo-organic compounds are isoelectronic with their carbon analogues. In $B_3N_3H_6$ the three non-bonding pairs on the nitrogen atoms behave like benzene and form an aromatic sextet.

B.5(4a) GROUP M4

Carbon is much the most electronegative element in the Group; but the electronegativities of the remaining elements are of similar magnitudes but of irregular order: thus $C \rangle Ge \rangle Pb \rangle Sn \rangle Si$. The ionisation energies of the elements are rather similar, tin being the lowest. The relative electrode potentials of tin and lead are almost the same. Under these circumstances it is not surprising that chemical properties show few regularities; indeed the only well defined Group trend is the increasing length of the covalent bond between identical M4 atoms.

The change from non-metallic to metallic lattice in the elements is probably caused by this increasing weakness of the M—M bond as the elements become larger. Tin can occur both as a covalent and a metallic lattice, each of which is stable under different conditions (as Napoleon found on the retreat from Moscow when the buttons on his soldiers' jackets turned into a non-metallic powder). The energies of the two forms are very nearly the same, so a very small change can result in two completely different structures—little wonder that the hard and fast classification into metal/non-metal is unreal when applied to the M4 elements.

The strength of the Ge—Ge bond is less than that of the Si—Si bond, but the resistance of the germanium chains to hydrolysis by alkali is very much greater. This is readily explained in terms of the greater electronegativity of Ge. In the Si—H bond the partial positive charge is on the silicon atom, rendering it liable to attack by OH^-, but in the Ge—H bond the germanium atom carries a partial negative charge which repels the hydroxide ion. Again, the bond strength of Ge—Cl will be lower than that of Si—Cl, so that it will be more readily reduced by Zn/HCl to GeH_4.

The comparatively low electronegativity of silicon will lead to violent reactions with the halogens and very high bond strengths. The Si—O bond is extremely strong (p. 36) since it has partial multiple bond character owing to the incorporation of the non-bonding electron pairs on the oxygen into the empty $3d$ orbitals on the silicon. Thus the halides of silicon, strong though the Si—Hal bonds may be, are hydrolysed by water, and so are the substituted halides like $(CH_3)_2SiCl_2$. However, the strength of the M—O bond decreases down the Group, so halide hydrolysis becomes less favourable. The empty $3d$ orbitals on silicon will allow this element to act as an electron acceptor with F^- forming the stable $SiF_6{}^{2-}$ ion, which accounts for the ability of anhydrous HF to attack silica glass.

The environment of all the fluorine atoms in the lattices of SnF_4 and PbF_4 is not identical; on account of the partial negative charge which they carry, some of them are attracted to more than one metal atom, giving a three dimensional structure which is consequently involatile.

The increasing basicity of the dioxides down the Group is a consequence of several factors, among them the relative falling off in the strength of the M—O bond; this is also the reason why the silicate chemistry is so much more extensive than that of the equivalent compounds of the lower members of M4. The formation of the octahedral $X(OH)_6^{2-}$ ions by tin and lead is not surprising in view of their larger size, and there is a parallel in the chemistry of antimony(v) and tellurium(vi).

The decreasing strength of the covalent bonds in the tetraethyls is accounted for by the increasing sizes of the atoms of the M4 elements.

The factors which contribute to the relative stabilities of the oxidation states of $+2$ and $+4$, and whether the former will be predominantly ionic or covalent, have already been discussed (p. 120). The dominant features of the chemistry of tin(II) are its reducing power, and its tendency to combine with the oxide or hydroxide ion, both of which are polarisable and can donate electrons to the tin. $SnCl_2.H_2O$ has a pyramidal shape for the same reason as NH_3, for it also has one non-bonding pair of electrons which are directed towards the fourth apex of a tetrahedron.

In its salts with anions which cannot readily donate electron pairs, the chemistry of Pb^{2+} resembles that of Sr^{2+} which is almost the same size. However, if the anions are polarisable they will donate electrons to the unfilled p orbitals of the lead; thus the sulphide is insoluble in acid, the insolubility of the halides increases from the chloride to the iodide, and many anionic complexes are formed. The H_2O molecule is not a very strong donor, and the charge/radius ratio for lead is only moderate, which means that the heat of hydration of lead salts is not high. However, like Sn^{2+}, Pb^{2+} will combine strongly with the oxide anion, which leads to extensive hydrolysis in solution and a tendency to precipitate basic salts.

B.5(4b) CARBON

As usual in the chemistry of elements of the second Period, their unique properties depend upon their small size, and their consequent inability to use the $3d$ orbitals for bonding.

The ability of carbon to catenate depends upon the following factors:

(1) The short bond distance and consequent high strength of the C—C bond means that carbon chains are thermally stable.
(2) Despite the high strength of the C—C bond, the formation

$$\diagdown\!\!\diagup C\!\!=\!\!O \quad \text{from} \quad C\big\langle\substack{C-\\ \\C-}$$

is energetically favourable even at low temperatures; yet the lack of feasible mechanisms with low activation energy renders carbon chains very resistant to oxidation below 300°C. In contrast, the lower strength of the Si—Si bond, the relatively high strength of the Si—O bond, and the available 3d orbitals on the silicon atom which can bond with incoming oxygen atoms all combine to render chains of silicon atoms extremely prone to oxidation. It is not surprising therefore that the higher silanes are pyrophoric.

Neither of the following two isoelectronic series of compounds have analogues in the chemistry of the other M4 elements:

| C=CH$_2$ | C=NH | C=O |
| C≡CH | C≡N | C≡O |

The reason is that the overlap of the atomic p orbitals to form stable multiple bonds is not possible except for the very small atoms that occur in the second Period.

Carbonium ions, carbon free radicals and carbanions are the more stable the greater the number of atoms over which the charge can be delocalised (p. 102) by alternative electron distributions. The triphenylmethyl ion structure for example.

Once again this particular mechanism is unavailable for the other M4 elements which cannot form this type of multiple bond.

There is bound to be some covalent character in the bonding of the C^{4-} ion in Al_4C_3: however the strongly polarising Al^{3+} ion has high deforming power (p. 99) and favours the small monatomic C^{4-} as opposed C_2^{2-}. The effect is exactly analogous to the formation of M_2O compounds by the smaller M1 metals, but M_2O_2 or even MO_2 by the larger ones (p. 117).

The compounds of carbon with the transitional elements are interstitial; they are formed very easily owing to the small size of the carbon atom.

The decrease in thermal stability of the halides of carbon from CF_4 to CI_4 is expected as the covalent bond becomes longer (p. 104). Their extreme resistance to hydrolysis—in sharp contrast to $Si(Hal)_4$—is a result of the inability of carbon to contain more than four atoms in its co-ordination sphere, owing to the unavailability of the d orbitals which in the case of silicon allow the intermediate formation of $SiCl_4(OH)_2$. $CHCl_3$ is more easily oxidised than CCl_4 because the high partial positive charge on the carbon weakens the C—H bond. The extreme stability of the fluorocarbons is probably the result of electrostatic attraction between the fluorine atom and the carbon adjacent to that to which it is bonded (p. 140).

B.5(5a) GROUP M5

The angles between the bonds of the tetrahedral X_4 allotropes is only sixty degrees, which involves a considerable amount of strain owing to electron repulsion. The X—X bond strength decreases down the Group, which means that the X_4 allotropes become increasingly unstable with respect to the continuous three dimensional structures.

The binary compounds have just those properties that would be expected of elements intermediate between metals and non-metals. The electron affinity of phosphorus is just high enough to prevent the P^{3-} ion being distorted by losing its charge in the crystal lattice of the phosphides of electropositive metals.

The thermal instability and reducing power of the hydrides and alkyls follows from the weakening of the covalent bonds to carbon and hydrogen with the increasing size of the M5 atoms. The same factor operates to make the XH_4^+ ion increasingly unstable: $(C_6H_5)_4As^+$ exists however. The weakening of the X—X bond for the same reason makes the H_2X—XH_2 compounds increasingly rare.

The weakening of the bonds of the M5 elements with halogens owing to the increasing size of the former dominates the strengthening effect caused by the increasing polarity of the bonds as the atoms down the Group become less electronegative. Thus PF_5 is stable with respect to PF_3 and F_2, whereas the reverse is true for the fluorides of bismuth. PF_5 is naturally more stable than PCl_5, and PI_5 cannot exist because there is not room for five iodine atoms to pack round one small phosphorus atom. The penta-iodides of the later elements cannot exist because of the oxidising power of the Group oxidation state. The comparative involatility of PCl_5 is probably caused by a tendency towards an ionic structure: $PCl_4^+PCl_6^-$.

The stability of the P—O bond is very great, since like the Si—O bond it can attain partial double bond character by incorporating the non-bonding electron pairs on the oxygen atom into the empty $3d$ orbitals on the phosphorus atom. The strength of the element-oxygen bond decreases down the Group, which renders the oxides of the later members of M5 more liable to attack by acids, and favours the formation of the lower oxide rather than the higher. The formation of the hydrated Group oxide, rather than the nitrate, when the element is dissolved in concentrated nitric acid is characteristic of the less electronegative non-metals and metalloids, for the partial negative charge of the oxygen bound to the metalloid is not high enough to stabilise the nitrate anion.

Owing to the anomalously high electronegativity of arsenic, the oxides and oxyacids of this element in the Group oxidation state are stronger oxidising agents than expected (p. 40).

The structure of P_4O_{10} is derived from the P_4 tetrahedron, with oxygen atoms in each edge and at the apices. The partially positively charged hydrogen of a water molecule attacks one of the partially negatively charged oxygens on the edge of the tetrahedron, and the partially negatively charged oxygen atom in the water attacks an adjacent partially positively charged phosphorus atom. As the temperature is raised further hydrolysis takes place until all the P—O—P links are broken and only PO_4^{3-} remains.

The lower oxyacids of phosphorus, dibasic H—PO—$(OH)_2$ and monobasic H_2—PO—OH are strong reducing agents, but $As(OH)_3$

has only limited reducing powers. H_3PO_3 is a weaker acid than H_3PO_4 because there are fewer oxygen atoms in the anion over which the partial negative charge can be spread, so that protons are more easily attracted.

The sulphides of arsenic and antimony are acidic, and will redissolve in any source of sulphide ions (inevitably alkaline) to give thio-arsenates and thioantimonates.

As the strength of the covalent bonds formed by the M5 elements falls down the Group, the formation of cations becomes energetically more favourable, until finally with Bi the cation Bi^{3+} is unambiguously formed. However, salts of Bi^{3+} are very easily hydrolysed, owing to the electron acceptor properties of the metal which can form a strong bond with the unshared pairs on the polarisable oxide anion. Thus $BiCl_3$ is stable in solution of HCl, but the insoluble BiOCl is precipitated on dilution.

PF_3 and PR_3, where R represents an alkyl or aryl group, form very strong complexes with transitional metals, since not only can they donate an electron pair in the same way as NH_3 (though not as strongly), but also they can accept an electron pair from the metal atom into their empty $3d$ orbitals. Nitrogen in ammonia has no $2d$ orbital in which to accept electrons, so amine and phosphine complexes are not strictly comparable.

The halides of M5 elements can accept halide anions to form anionic halide complexes, and so qualify as Lewis Acids (p. 104). PF_6^- is very stable.

B.5(5b) NITROGEN

The N—H bond is shorter, stronger and more polar than the hydride links of the other M5 elements, so much so that the boiling point of NH_3 is raised by hydrogen bonding (p. 135). Because of the larger partial negative charge on the nitrogen, ammonia will be a better donor to protons than phosphine, and H_2N—NH_2 will form salts whereas H_2P—PH_2 will not do so (p. 101).

By Fajans' Rules, N^{3-} is more likely to be stable than the other anions of the M5 elements, since it is smaller than they are and less polarisable.

Nitrogen does not have orbitals available for forming five covalent bonds. For this reason the co-ordination of a water molecule to the NCl_3 molecule cannot take place, as it does in the first stage of the hydrolysis of PCl_3, so it is not surprising that the products of these formally similar reactions are not analogous. The explosive properties of the nitrogen trihalides are accounted for by the relative weakness of the N—Hal bond, and the very great stability of molecular nitrogen.

Because of its small atomic size, nitrogen forms multiple bonds with carbon, itself and oxygen by the overlap of atomic p orbitals (p. 103). The strongest of these bonds is the N≡N in molecular nitrogen, and its stability dominates the chemistry of the element; for any compound for which there is a mechanism for decomposition in which N_2 is formed will be unstable, and almost certainly will be formed endothermically from its elements. For instance the nitrogen-oxygen bond in NO is very strong and this compound is hardly dissociated at 2000°C, yet this does not prevent reduction when the gas is passed over hot copper. NO is the most stable of the oxides of nitrogen, having six bonding and one anti-bonding electron, giving it a bond order of $2\frac{1}{2}$.

The oxyacids of nitrogen are only similar to the oxyacids of phosphorus in the Group oxidation states of the N and P; their formulae, structure and reactions are entirely different. The cause is the absence of $2d$ orbitals in nitrogen, which would allow the element to make use of more than eight valence electrons in bonding. Thus in the compound R_3NO, the nitrogen donates electrons to the oxygen atom; the same process takes place in R_3PO, but the oxygen can share its non-bonding electron pairs with the $3d$ orbitals of the phosphorus atom, so that the P—O link has a bond order greater than one, and is correspondingly stable and difficult to reduce. For the same reason phosphoric acid is also difficult to reduce too, whereas nitric acid, with the possibility of forming stable NO or NO_2 as end products (which phosphoric acid cannot do as PO and PO_2 do not exist) will be reduced by compounds normally considered to have only feeble reducing powers.

Since nitrogen is about as electronegative as oxygen, the partial charge on the oxygen atom in NO_3^- will not be high, so nitric acid will readily expel a proton and behaves as a strong acid. Nitrous acid is a weaker acid, for there is one less oxygen atom in NO_2^- as compared to NO_3^- over which to share the partial negative charge, so protons are more attracted to the former (p. 101). Nitrous acid is

unstable in concentration of greater than 0·2M, for it disproportionates to nitric acid and nitric oxide:

$$3HN^{III}O_2 \rightarrow HN^{V}O_3 + 2N^{II}O + H_2O$$

Unlike the nitrogen atom in the ammonia molecule, the nitrogen in
NF_3 carries a partial positive charge, so its power to donate an
electron pair is very small. PF_3 on the other hand can accept electron
pairs from metals late in the transitional Periods into its empty $3d$
orbitals. Thus $Ni(PF_3)_4$ is stable and analogous to $Ni(CO)_4$; in both
cases the oxidation state of the central metal atom is zero.

B.5(6) GROUP M6: OXYGEN AND SULPHUR

In terms of molecular orbitals, oxygen has six paired bonding and two
unpaired anti-bonding electrons, which account for its strong
paramagnetism. An alternative formulation of the O_2 structure by
Linnett relies upon 7 electrons of clockwise spin arranged at the
apices of two tetrahedra linked at one corner (with the oxygen atoms
at the centre of the tetrahedra) together with 5 electrons of
anticlockwise spin arranged at the apices of two tetrahedra linked at
three corners. Such a formulation minimises interelectronic repulsion
and also accounts for the paramagnetism by having two unpaired
electrons. Both the molecular orbital hypothesis and Linnett's
Double Quartet hypothesis fit the facts and are of predictive value,
but neither of them can be directly validated by an experimental
test of the positions of the various electrons (p. 77). No one who
enquires which of them is 'the truer' or 'the better' hypothesis has
any real understanding of the nature of theoretical science.

Sulphur is reluctant to form multiple bonds by the overlap of atomic
p orbitals, though it probably does so in a few compounds,
S=C=S for example. In this way the pair of single bonds which
allow chain formation are stabilised with respect to the multiple
bond which would lead to S_2 molecules. The demonstrable weakness
of the —O—O— bond (similar to the weakness of $>$N—N$<$ and
F—F) is probably caused by the repulsion of the non-bonding pairs
on the two oxygen atoms. Certainly the structure of H—O—O—H
(the shape of the molecule can be visualised with the —O—O— bond
up the spine of a half opened book and the two OH groups in the
plane of the pages to maximise the distance apart of the electron
pairs) bears out this hypothesis. In contrast, the —S—S— bond is
longer than the —O—O— bond, and so is not weakened by the
repulsion of the electron pairs on adjacent atoms in the latter. The

—Se—Se— bond is longer than the —S—S— bond, and so as usual it is weaker, thus catenation is less important in the chemistry of selenium than it is in the chemistry of sulphur.

Oxygen is an extremely electronegative element, so the removal of electrons in the formation of oxygen fluorides is difficult, and the compounds formed are unstable. The sulphur-oxygen bond is very strong for the same reason that Si—O (p. 129) and P—O (p. 131) are strong; thus SO_2 is oxidised to SO_3, and SO is so readily oxidised that it does not exist. SO_3 is a tremendously strong Lewis acid, accepting electrons from HO_2, H_2SO_4, and even the unshared pairs of chlorine in the covalent HCl molecule.

The H—O bond is considerably shorter, and therefore stronger, than the H—S bond, which explains the lower thermal stability of the latter. Water will accept protons from other H_2O molecules to the extent of 1 part in 10^7; when H_2S is dissolved in water the weaker H—S bond will break more easily and H_2S is dissociated to the extent of 1 part in 10^4.

By extrapolating the boiling points of the other hydrides of M5, M6 and M7, the figures for NH_3, H_2O and HF should be respectively about $-110°C$, $-80°C$ and $-105°C$, whereas in fact they are $-33°C$, $100°C$ and $20°C$. This apparent anomaly is explained in terms of **hydrogen bonding**—or, more accurately, **protonic bridging**—in which the partially positively charged hydrogen can, by virtue of its very small size, be attached to two atoms if they are sufficiently electronegative. In this way association takes place, and the actual molecular weights of the hydrides of N, O and F are very much greater than is suggested by the simple formulae. Nitrogen is the least electronegative, so N—H—N interactions will be the weakest. The F—H—F⁻ ion is symmetrical and occurs in crystals of KHF_2;

long chains of 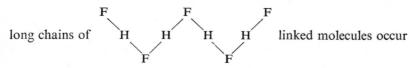 linked molecules occur

in anhydrous HF, and explain its strongly acid nature (p. 140). H_2O, with two hydrogen atoms which can form strong protonic bridging, can associate in three dimensions, which is why its boiling point is the highest of the three. Hydrogen bonds have strengths of 5-20 kJ mol⁻¹, and are intermediate in strength between the weakest chemical bonds like F—F (159 kJ mol⁻¹) and the strongest van der Waals interactions (1-2 kJ mol⁻¹).

For the oxides and sulphides of M1, the heat of hydration is always greater than the lattice energy, and hydrolysis is always complete. For M2 elements the hydroxides are formed irreversibly from the oxides and water, but their lattice energy is high, and they are not very soluble; solubility inceases down the Group.
The lattice energy of the sulphides is not so high and their solution is therefore complete—especially on warming, which causes H_2S to be expelled from the liquid.

The heats of formation of oxides, which are the more polar, are always higher than those of the corresponding sulphides. This is why the metallurgy of all elements occurring in nature as sulphide ores depends upon roasting them to the oxide before reducing the latter, for the reaction:

$$MS + 1\tfrac{1}{2}O_2 \rightarrow MO + SO_2$$

is very favourable under these conditions.

However, among the metals and metalloids of the later Main Groups, the sharing of the non-bonding electron pairs on the sulphur atom with these elements of high electron affinity leads to extra stability for the sulphide lattice, so the lattice energies of the oxides and sulphides of these elements differ very little. Oxides always dissolve in acids by a process which can be described by the equation:

$$MO + 2H^+(aq) \rightarrow M^{2+}(aq) + H_2O$$

because the high heat of formation of water, and the high heat of hydration of metal ions, will always provide sufficient energy to disrupt the oxide lattice. However, the production of H_2S, which has a lower heat of formation, by an analogous reaction is not nearly so energetically favourable, so sulphides with a high lattice energy may very well be insoluble in acids.

The non-stoichiometric oxides and sulphides of transitional metals are generally defect lattices in which some of the metal atoms are missing, while the others show a higher oxidation state: thus $Fe_{0.89}S$ contains some $Fe(\text{III})$ atoms.

The hydroxides of amphoteric elements of intermediate electronegativity will often take up hydroxide ions and eliminate water with the formation of oxyanions. Sulphides of some later Main Group metals will under alkaline conditions take up further sulphide ions to form compounds in which the metal is on the anion.

Thiostannates and thioarsenates are examples of this behaviour, which is a consequence of the stability of the metal–sulphur bond.

The stability of the X—H bond in M6 falls down the Group, which makes them thermally less stable but stronger acids. Their polarity is intermediate between the mainly basic M5 hydrides and the acidic M7 hydrides.

The oxides are typically acidic, in that they dissolve in water and are prepared by oxidising the element with concentrated nitric acid; but as the polarity of the X—O bond increases down the Group, their acidic properties become less pronounced.

The oxyanions of the Group oxidation states of arsenic, selenium and bromine are all anomalously strong oxidising agents (p. 107), and they do not form polymeric oxyanions owing to the weakness of their bonds with oxygen. The oxyacid of tellurium is notably weaker than sulphuric and selenic acid owing to the relatively high partial negative charge on oxygen when bonded to tellurium, the most electropositive member of the Group.

The increase in reactivity from SF_6 to SeF_6 and TeF_6 is caused by the weakening of the X—F bond and the less efficient shielding of the larger atoms by fluorine. In addition, elements in the later Periods can use more than 12 electrons in their bonding molecular orbitals.

Cations are most favoured in the elements of high atomic weight in the later Main Groups, where the strength of the covalent bonds has fallen owing to the increasing size of the elements. However, the later the Main Group, the rarer it is for cations to be formed, since the ionisation energy of the slightly penetrating $5p$ and $6p$ electrons rises across the Period with the increase in nuclear charge.

B.5(7a) GROUP M7: THE HALOGENS

Chlorine is more electronegative than bromine which in turn is more electronegative than iodine; thus the bonds between chlorine and the less electronegative elements will be stronger than those of the other two elements—because they are both shorter and more polar, chlorine being smaller. The high bond strength of bonds containing chlorine explains why chlorine will bring out higher oxidation states in the elements with which it combines than will iodine, for more energy is available for the 'promotion' of the necessary electrons.

With electropositive elements, three dimensional co-ordination polymers with regular crystal structures are formed, each halogen being attracted equally to 6 or 8 metal atoms and vice versa. Such compounds are salts, because at high temperatures or in solution the potential ions in the co-ordination polymer actually become free and will therefore conduct electricity. The fluorides form the most polar bonds, and they are still high melting solids with 6 co-ordination which usually sublime when the other halide compounds have passed over into weakly bonded crystals with 4 co-ordination (see A.5(4)).

The halides of the less electropositive metals often have covalently bonded molecules—the dimeric Fe_2Cl_6 for example. These react with water to form hydrated cations, which then hydrolyse by expelling a proton from the water molecules co-ordinated round the central metal atom.

Compounds of the non-metals with the halogens (not fluorine) are usually hydrolysed by water, since the strength of the non-metal oxygen bond will be greater than that of the non-metal halogen bond owing to the greater electronegativity of the oxygen atom. Water will donate its non-bonding electron pairs to the d orbitals of the non-metal, and the halogen atoms bonded to it will be eliminated as halogen hydride.

With the metals of M1 and M2 the strongest bonds are formed by fluorine because of its higher electronegativity; thus the lattice energies of fluorides will be higher than those of the other halides, and fluorides will be more insoluble. In the later Main Groups, where the donation of electrons to the metal atom increases the lattice energy, iodide ions are the best donors, so iodides have the highest lattice energies and are the most insoluble halides. The donor properties of the iodide ion also account for the fact that anionic iodide complexes are the most stable in the later Main Groups.

The bond strength of the hydrogen halides decreases down the Group (longer bond, lower polarity), so HI is much more difficult to form from its elements than HCl and it dissociates more easily. For this reason it is the strongest acid and the strongest reducing agent; iodides metals in their of higher oxidation states tend to be unstable.

The difference in electronegativity between iodine and oxygen is greater than it is for the other halogens, so it is not surprising that

I_2O_5 is the most stable halogen oxide. All the oxides and oxysalts will form halogens or halide ions on heating; the affinity of bromine for oxygen is anomalously low (p. 107). Le Chatelier's Principle operates to increase the solubility of the halogens in alkali by the removal of protons. The unexpected stability and weak oxidising power of the ClO_4^- ion is probably caused by its symmetrical structure and very low polarisability. The octahedral structure of the periodates is not surprising since the elements in Period 5 are large enough to favour 6 co-ordination.

Interhalogen compounds containing more than two atoms always contain an excess of the electronegative element. These compounds will always disproportionate on heating because the drive towards disorder is favoured by a rise in temperature, thus:

$$ICl_3 \rightleftharpoons ICl + Cl_2$$

The polyhalide ions are generally only stable in combination with large non-polarising cations like Cs^+, and any attempt to form such salts with more polarising cations leads to decomposition.

Because of its relatively low ionisation energy, iodine is far more likely to form cations than the other halogens. In carrier experiments astatine has also been shown to form a stable At^+ ion.

B.5(7b) FLUORINE

The reactivity of organic iodides, and the reluctance of iodine to add on to carbon-carbon double bonds, result from the large size and less electronegative nature of the iodine atom which consequently leads to a low carbon-iodine bond strength.

Because of its great electronegativity and very small size, fluorine forms strong bonds with other elements, including krypton and xenon. The extremely great heats of reaction where fluorine is involved are also the result of the very low dissociation energy of the F_2 molecule, which is thought to be caused by the repulsion of the non-bonding pairs (p. 103). The great strength of the bonds formed accounts for the power of fluorine to bring out the highest oxidation state of any element with which it will combine.

The remarkable inertness of the octahedral molecule SF_6 (used as an insulator in electrical switchgear) depends partly on the strength of the S—F bond, partly on the fact that the six peripheral fluorine atoms effectively prevent any other atom from attacking the sulphur and partly because there are no low energy orbitals available to which an approaching atom could donate a non-bonding electron pair (as water does in the hydrolysis of $SiCl_4$). The contrasting reactivity of UF_6 depends upon the large size of the uranium atom, which can readily accommodate sixteen electrons in the co-ordination sphere.

As expected, fluorides of very electropositive metals have higher lattice energies than the other halides so melt at higher temperatures, while fluorides of electronegative elements have smaller molecules than the other halides so the van der Waals forces are smaller and they are more volatile. The fluoride of a metal in a low oxidation state is always more polar—and therefore more involatile and thermally stable—than the fluoride of the same metal in a higher oxidation state.

Anhydrous hydrogen fluoride is polymerised in the liquid state by hydrogen bonding (p. 135), and this accounts for the very strong proton donor qualities of this substance: it will displace HCl from metallic chlorides and form $H_2NO_3^+$ with nitric acid. When it ionises the negative charge can be dispersed over a large anion which is more favourable than confining it to a small HF_2^- ion. Thus:

$$(HF)_{n+1} \leftrightharpoons H^+ + H_nF_{n+1}^-$$

when anhydrous and acting as an acid. The very slight conductivity of anhydrous HF is caused by the ionisation into H_2F^+ and $H_nF_{n+1}^-$, HF reluctantly acting as a very weak base. Metallic fluorides act as bases in HF, whereas very powerful fluoride acceptors like SbF_5 will act as acids: e.g. $H_2F^+SbF_6^-$.

The polymeric anions in anhydrous HF are broken up when the acid is diluted, and the resulting solution is only 1% ionised according to the equation:

$$H_2O + 2HF \rightleftharpoons H_3O^+ + FHF^-$$

This reluctance to ionise is caused by the very high bond strength of the H—F link, for compared to the other hydrogen halides the extra heat of hydration of the small fluoride ion is cancelled out by the ordering effect it has on the surrounding water molecules.

Since fluorine is more electronegative than oxygen, it will not donate electrons to the latter which would be necessary if it were to form oxyacids analogous to those of chlorine.

The very small bond distance in the C—F link means that fluorine atoms can be attracted electrostatically to carbon atoms adjacent to those with which it forms normal covalent bonds; this attraction

will add to the carbon-fluorine bond strength and hinder the rotation
of the carbon chain. This explanation is purely hypothetical, but it is
supported by the high viscosity of liquid fluorocarbons, a natural
consequence of hindered free rotation.

B.5(8) GROUP M8: THE NOBLE GASES

All the noble gases have accumulated in the atmosphere except
helium and radon: the former because it is not held by the
gravitational field of the earth, the latter because it is radioactive
and has a short half-life. These elements were discovered in 1895 by
Lord Rayleigh, as a result of a discrepancy between the molecular
weight of 'atmospheric nitrogen' and nitrogen prepared by purely
chemical means. The emission spectrum of helium had been detected
in the solar spectrum 30 years before.

No true chemical compounds of these elements were reported for a
further 60 years, though hydrates of argon, krypton and xenon
(dissociating well below 0°C) were known and clathrate compounds
of argon and krypton—in which the monatomic gas molecules are
locked in the interstices of crystals of hydroquinone—can be formed
quite readily. Also there was spectral evidence for such molecules as
He_2^+ in discharge tubes. However, the fundamental unreactivity of
these elements was attributed to a rhythmic electron pattern of
exceptional stability, and this assumption was very valuable in
promoting the growth of modern theories of periodicity (p. 81).

The very great danger of treating a provisional scientific hypothesis
as an article of faith is wonderfully well illustrated by what happened
in this case. Students all over the world were taught that the
hypothetical unreactivity of the noble gases had the quality of
revealed truth, and spent their time manipulating elements adjacent
to the noble gases in the Periodic Table so that they might 'gain or
lose electrons' (p. 97) to produce anions and cations of the same
inviolable stability. Little wonder that the absence of compounds of
the noble gases could never be questioned, even though experimental
evidence of measured ionisation energies made it embarrassingly
clear that xenon ought to have a chemistry. Plenty of compounds of
O_2^+ with large non-polarising anions like PtF_6^- were known by
1962, and the reported ionisation energy of Xe was lower than that of
O_2. It was left to Bartlett to fly in the face of convention, and having
successfully formed O_2PtF_6 by mixing the gases O_2 and PtF_6 at room
temperature, to repeat the experiment substituting xenon for oxygen.

The bond energy of the Xe—F bond in the xenon fluorides is approximately 138 kJ mol^{-1}, so the compounds are thermodynamically stable, being formed exothermically from their elements. The oxygen compounds are less stable, but this reflects the high heat of atomisation of the oxygen molecule (493 kJ mol^{-1}) as compared to the fluorine molecule (154 kJ mol^{-1}) rather than a greater weakness in the Xe—O as opposed to the Xe—F bonds. The Kr—F bond is much weaker than either of these.

The X-ray spectra of XeO_3 and XeO_4 are very similar to the isoelectronic IO_3^- and IO_4^-, which suggests that they are respectively pyramidal and tetrahedral. XeF_2 is linear and XeF_4 is square planar; it is not yet clear whether XeF_6 is a regular or distorted tetrahedron, but the abstruse measurement to determine this has important consequences for bonding theory.

Clearly bonding takes place by a substantial transfer of electrons from the xenon to the electronegative element. Odd numbered oxidation states are not found, for these would involve an unpaired electron remaining on the noble gas atom. The bonds will necessarily be stronger the lower the ionisation energy of the noble gas and the more electronegative the peripheral atoms. Thus compounds of xenon and radon with fluorine and oxygen should be the most stable, whereas the stability of the krypton fluorides and radon chlorides should only be marginal.

B.5(9) TRANSITIONAL ELEMENTS OF THE FIRST LONG PERIOD

In the first transitional series, the completed rhythmic patterns of orbitals which make up the argon core act as a shield to the non-penetrating 3d electrons, the probability distribution for which lies mostly outside that of the 3s and 3p orbitals. Thus the 3d orbitals are in a position to interact with the environment, and lots of interesting chemistry will result.

In the discussion of the first ionisation energy of lithium (p. 85), it was clear that the two 1s electrons shielded the 2s electron from a nuclear charge of +3 very efficiently except for that small part of the 2s probability distribution which penetrated close to the nucleus. Consider now the successive addition of protons (with their appropriate neutrons) to an argon core with its 18 extra-nuclear electrons. The electrons in the 4s orbitals, which are penetrating, will feel an extra attraction which rises in proportion as the nuclear charge increases from $18 \rightarrow 19 \rightarrow 20 \rightarrow 21$ etc. However, electrons in the $3d$ orbital, which are non-penetrating, will feel an increase of nuclear charge in proportion to $1 - 2 - 3 - 4$ etc., as successive protons are added to the nucleus. Thus the $4s$ orbitals are stabilised gradually as the number of protons in the nucleus increases, whereas the stability of the $3d$ electrons, rather low at first, rises very sharply for each addition of a positive charge. When there are two positive charges in excess of the argon nucleus, the $4s$ orbital still is more stable than the $3d$, but with the addition of one further positive charge, the $3d$ orbital becomes more stable than the $4s$. Thus the normal sequence of ionisation for titanium, the fourth element after argon, is as follows:

argon/$3d^2 4s^2$	argon/$3d^2 4s^1$	argon/$3d^2$	argon/$3d^1$	argon
Ti	Ti$^+$	Ti^{2+}	Ti^{3+}	Ti^{4+}

(a) Since there are many more vacancies of low energy than there are valence electrons in the transitional elements, they will all show metallic bonding. The density of the metals will increase across the Period, for the additional nuclear charge will attract the electron cloud increasingly strongly and cause it to contract. The ionisation energy of the $4s^2$ electrons will rise and so the transitional metals become increasingly less electropositive across the Period.

Transitional metal cations are doubly or triply charged and tend to become smaller across the Period. They will therefore be highly polarising, and will tend to have properties rather like Zn^{2+} or Al^{3+}: their salts will be hydrated and extensively hydrolysed in solution, and the anhydrous chlorides will in some cases sublime.

(b) The change in size from one similarly charged ion to the next across the Period is not large, and therefore it is not surprising that compounds of similar crystal structure are formed by successive elements.

(c) The difference in energy given out when compounds of oxidation state $+n$ and $+(n + 1)$ are formed is generally not very great for the transitional elements. Up to manganese they can (but need not) use all the electrons outside the argon core for bonding; but in later elements the nuclear attraction becomes too strong for this to happen.

The universal formation of M^{2+} by the transitional elements corresponds to the removal of the two $4s$ electrons common to them all. Since they are increasingly strongly held across the Period, the standard electrode potential of this oxidation state will fall from scandium to copper. The M^{2+} ions early in the Period are unstable with respect to conversion to higher oxidation states, but the oxidation state of $+2$ becomes increasingly stable across the Period as it becomes more difficult to ionise further electrons.

(d) In a non-homogeneous chemical environment not all the d orbitals have the same energy. Transitions of electrons between these different levels will absorb light over a band of the visible region of the spectrum.

(e) Homogeneous catalysis depends upon the alternate oxidation and reduction of a metal ion, and the transitional metals, with small differences in energy between their various oxidation states, can readily act in this way.

Heterogeneous catalysis is a surface effect which depends on having low energy orbitals on the surface of the catalytic material which can interact with molecules as they are adsorbed and weaken their bonds. Transitional metals have many such low energy orbitals available.

(f) Paramagnetism is found for any compound which contains unpaired electrons. Since in the free atoms Hund's Rule demands that all the d orbitals are half filled before any of them are double occupied, any compounds (but not tightly bound complexes) which contain from one to nine d electrons will be paramagnetic.

The anomalously high electrode potential of Mn^{2+}/Mn, and the relative stability of Fe^{3+} as compared with Fe^{2+} depend upon the unusual stability of the $3d^5$ configuration, in which the five $3d$ orbitals are all half filled.

(g) Transitional metals and ions have plenty of low energy orbitals available which can combine with suitable orbitals on incoming ligands to form very stable bonding molecular orbitals.

Consider a transitional metal at the centre of a cube. Two $3d$ orbitals of complex shape will be directed to the centres of the six faces: these are called the e_g orbitals. Three further $3d$ orbitals, called t_{2g} orbitals, are directed at the 12 mid-points of the edges of the cube. The formation of an octahedral complex involves molecular orbitals made by combining filled ligand orbitals with empty e_g orbitals; any d electrons previously in the e_g orbitals are promoted to the new anti-bonding molecular orbitals, and the more stable the complex formed, the higher in energy these anti-bonding molecular orbitals will be. The formation of a tetrahedral complex involves similar bonding with the t_{2g} orbitals.

If the anti-bonding molecular orbital of an octahedral complex is of high energy, the energy required to promote the unpaired d electron to it will be greater than that required to transfer it to a half filled t_{2g} orbital and pair it with the electron already there. Thus Fe^{3+}, with five unpaired electrons, will on the approach of six CN^- ions which are excellent donors, pair the two $3d$ electrons originally in the e_g orbitals with the single $3d$ electrons in two of the three non-bonding t_{2g} orbitals, so the resulting complex contains only one unpaired electron. This is a 'low spin complex' as opposed to $Fe(H_2O)_6^{3+}$ which is a 'high spin complex' with 5 unpaired electrons; in the latter the bonding molecular orbital is not very stable, so the energy required to promote the e_g electrons to the anti-bonding molecular orbital is not sufficient to make these electrons pair with those electrons already in the t_{2g} orbitals.

Some unfamiliar oxidation states can be stabilised by complex formation. Co^{3+} has six $3d$ electrons; $Co(NH_3)_6^{3+}$ has promoted the two electrons from the e_g orbitals and paired them with the two unpaired electrons in the t_{2g} orbitals; thus it has no unpaired electrons and no electrons in high energy anti-bonding orbitals. Thus it is more stable than $Co(NH_3)_6^{2+}$ which has to accommodate its extra electron in an e_g anti-bonding orbital. In high spin complexes, the hydrated cations for example, $Co(H_2O)_6^{2+}$ is much more stable than $Co(H_2O)_6^{3+}$.

Complexes where the oxidation state of the metal is zero are also formed, usually by neutral ligands. Besides $Ni(CO)_4$ there are $Fe(CO)_5$ and $Cr(CO)_6$: there are also the sandwich compounds $Fe(C_5H_5)_2$ and $Cr(C_6H_6)_2$. In these compounds all the low energy orbitals are filled.

B.5(10) SUMMARY OF ANOMALIES SHOWN BY FIRST MEMBERS OF MAIN GROUPS

The anomalies in the chemistry of the first elements in the Main Groups all stem from their very small size in relation to the later members. It is convenient to divide the factors that result from small size into two main groups; the first concerned with the effect of small size on bond strength, and the second concerned with the total number of bonds formed. This classification is as usual not mutually exclusive.

Ref. p. 148	M1 Li	M2 Be	M3 B
I(a)	Rather slowly attacked by water. Covalent higher alkyls.	Amphoteric oxide. Few salts, all basic.	No cations. Polymerised borates.
I(b)	High heat of hydration. Thermal decomposition of Li_2CO_3.	Chloride scarcely ionised.	
I(c)(i)	High lattice energy of LiF, Li_3PO_4.		
I(c)(ii)			
I(d)			
II(a)		Always 4 mols of water of crystallisation.	
II(b)	Involatile $LiCH_3$.	Polymeric alkyls. Bridging in $BeCl_2$.	Icosahedral structure of B. Electron deficient hydrides.
II(c)			Back co-ordination in BF_3.

M4 C	M5 N	M6 O	M7 F
No cations.	Hydrogen bonding in NH_3. Some N^{3-} formed.	Hydrogen bonding in H_2O.	Hydrogen bonding in HF. Stable F—H—F⁻. No oxyacids.
Thermal stability of carbon chains.			HF is reluctant to ionise. Elements show highest oxidation state in fluorides.
	N—N unstable.	—O—O— unstable.	F—F unstable.
C=C, —C≡C—, —C≡N, C=O, all exist.	N≡N, —N=N—, —N=O all exist.	C⁻=O⁺, C=O all exist.	
CCl_4 is not hydrolysed.	No NF_5.	No OF_6.	
No partial double bond character in C—O; thus no oxidation of carbon chains.	Anomalous hydrolysis of trihalides.		

I   Influence of small size on bond strength

(a) High electronegativity. All first members of Main Groups will have relatively high electronegativity compared with later members, for the valence electrons are closer to the nucleus and therefore more difficult to remove.

(b) High polarising power. The radius of first members is smaller than that of later members, so the charge/radius ratio will be higher and the ions will have greater deforming power.

(c) (i) Formation of strong bonds. First members of Main Group elements can approach each other very closely when combining together, leading to a short intermolecular distance and strong bonding.

(ii) In a few cases the repulsion on non-bonding pairs on atoms that have approached each other closely leads to bond weakening.

(d) Multiple bonding. Carbon, nitrogen and oxygen can, when forming single bonds, overlap their remaining atomic p orbitals with themselves or with each other to form multiple bonds of increased strength.

II   Influence of small size on number of bonds formed

(a) Limitation to four bonding orbitals. Lack of low energy orbitals prevents the formation of more than four covalent bonds by first members, for there are no $2d$ orbitals to be used, and the $3s$ and $3p$ orbitals are too high in energy.

(b) Delocalisation in electron deficient molecules. If there are two few electrons present to fill the four low energy orbital vacancies in first members, electron pairs will become delocalised and form three-centre bonds.

(c) Limited back-co-ordination. If 4 covalent bonds are already formed, they cannot attain partial multiple bonding by incorporating electron pairs from the ligands into their non-existent $2d$ orbitals.

The table on pp. 146-7 is only a summary and much has been omitted; the special sections under the Main Group headings should be consulted for a detailed consideration of any particular anomaly.

B.6
Special topics

B.6(1) DIAGONAL RELATIONSHIPS

The similarities between Li^+ and Mg^{2+} and between Be^{2+} and Al^{3+} can be readily understood in terms of polarisation, as defined by the ratio charge/radius. The second member of these two pairs has a higher charge, corresponding to its later Periodic Group, and a higher radius corresponding to its later Period, so the polarising power of both ions in a pair will be roughly the same—which in the case of electropositive metals leads to similar chemical behaviour.

Inevitably the unit cell of a crystal of LiX cannot be the same as a crystal of MgX_2. The ionisation energy of magnesium is too high for the formation of an ionic hydride, and the match of charge in the two ions Mg^{2+} and O^{2-} confers a high lattice energy on MgO with consequent insolubility.

Beryllium is too small either to accommodate six water molecules in its co-ordination sphere, or to carry two negative charges on the BeH_4^{2-} ion.

Boron has a similar electronegativity to that of silicon, and the bonds formed by these two non-metals with other elements are similar in strength. The very strong Lewis acid properties of BF_3 depend upon the unfilled orbital of low energy on the boron, but since it is in the first Period this element cannot accommodate more than eight valence electrons, so there is no exact analogue to SiF_6^{2-}.

B.6(2) CONTRAST OF THE FIRST TRANSITIONAL SERIES WITH THE SECOND AND THIRD TRANSITIONAL SERIES

Elements in the first transitional series readily lose the $4s^2$ electrons; but since these electrons are penetrating, the tendency to lose electrons decreases across the Period with the increase in nuclear charge. In the second and third transitional series the $5s^2$ and $6s^2$ electrons are still penetrating, so they are even more strongly held than the $4s^2$ electrons; therefore the electrode potential for the change $M^{2+} \rightarrow M$ is below that of hydrogen, and the elements are unreactive.

All transitional metals can form anionic complexes, for their empty orbitals can accept electrons from anions like Cl^-. In the first Period the relative stability of cations and anionic complexes is nicely balanced: $Co(H_2O)_6^{2+}$ (pink) will readily convert to $CoCl_4^{2-}$ (blue) with any solution containing chloride ions. In the second and third transitional Periods the available orbitals on the larger metal ions are lower lying, so more anions can be accommodated and they will be bound more strongly. The falling stability of cations, and the rising stability of these anionic complexes in the second and third transitional Periods is bound to lead to a preponderance of the latter.

Although the ionisation energy for the $5s^2$ and $6s^2$ electrons is high, that for the successive $4d$ and $5d$ electrons is rather lower than for the $3d$ electrons. Thus the energy gained from the formation of extra bonds is more than enough to promote the d electrons into bonding molecular orbitals, with the consequent formation of compounds in high oxidation states, very often with the transitional metal of the second or third Period contained in the anion.

The stability of these high oxidation states extends to the oxides, especially if the molecule so formed is a simple symmetrical one like OsO_4, where osmium is a large enough atom to contain 16 valence electrons in its bonding molecular orbitals.

B.6(3) SIMILARITY OF SECOND AND THIRD TRANSITIONAL SERIES: THE LANTHANIDE CONTRACTION

The similarity of the second and third transitional series, particularly at the beginning, is the result of the existence of the first Period of inner transitional elements. The latter, as expected, show a steady decrease in the radius of their atoms and ions as the nuclear charge increases; thus the radius of La^{3+} is about 0·2 Angstroms larger than that of Lu^{3+}. This makes the third transitional series 0·2 Angstroms smaller than expected, a new factor which cancels almost exactly the theoretical increase in size to be expected in the descent of any Group. Thus the atoms of Zr and Hf in their analogous compounds are almost exactly similar in size, which accounts for their extraordinarily similar properties.

The cause of the lanthanide contraction is the same as the cause of the contraction across the transitional Periods, namely the very imperfect shielding (p. 86) of one electron by another in an incomplete rhythmic pattern of orbitals. Across the inner transitional series the number of f electrons and the nuclear charge increase by one at each step; thus the thirteenth $4f$ electron is held by a nuclear charge not much less than thirteen times as great as the first $4f$ electron. Naturally this increase in coulombic attraction on all the $4f$ electrons present, and on those other electrons lying outside the $4f$ orbitals, causes a reduction in the effective size of the atom.

B.6(4) THE CHEMISTRY OF THE INNER TRANSITIONAL ELEMENTS: THE LANTHANIDES AND ACTINIDES

The electronic configuration of lanthanum is /xenon core/$6s^2.5d^1$. The shielding of the filled orbitals of the xenon core is very efficient, so the three peripheral electrons are easily ionised, leaving stable La^{3+} salts.

The shapes of the $4f$ orbitals are irregular, but two generalisations can be made. First, they do not penetrate close to the nucleus, and secondly their probability distribution hardly extends outside that of the $5p$ orbitals which are already filled, let alone the higher s and p orbitals. Thus the $4f$ orbitals are buried relatively deep within the atom and therefore their interactions with the environment are minimal, and not chemically significant. This is why the chemistry of the lanthanides is so homologous compared to the elements of the first transitional Period, in which the $3d$ orbitals project well out to the periphery of the atoms and ions where they can both donate and receive electron pairs.

The penetrating $6s^2$ electrons will be more strongly held across the lanthanide Period owing to the increased nuclear charge so that the relative electrode potential of lanthanum (element 57) will be above (more negative than) that of lutecium (element 71), just as that of calcium is above that of zinc. The removal of a fourth electron becomes increasingly difficult from cerium (element 58) to praesodymium (element 59).

The Hund's Rule prediction of tne extra stability of an empty, half full and full set of orbitals corresponding to a configuration of $4f^0$; $4f^7$; $4f^{14}$, is borne out by the existence of Ce^{4+}; Eu^{2+}; invariable Gd^{3+}; Tb^{4+}; Yb^{2+}. The fact that there are some other transiently stable oxidation states which do not have the stabilised configurations shows that Hund's Rule, like the noble gas rule for Main Group elements, expresses a tendency—not an absolute rule.

In the actinide Period the $5f$ orbitals are much closer in energy to the $6d$ orbitals than the $4f$ are to the $5d$ in the lanthanides: they are not so deeply buried, so they can take a much more active part in chemical bonding. This accounts for the greater prevalence of high oxidation states early in the Period, but these become less favourable as the nuclear charge builds up, until towards the end of the actinide Period the oxidation state of $+3$ is paramount. The greater number of available orbitals contributes towards the tighter bonding of co-ordinated water molecules with consequent hydrolysis in the fashion of $Al(H_2O)_6{}^{3+}$, and also to complex formation with other ligands which are electron pair donors.

Postscript

Is Chemistry only Applied Mathematics?

1 THE ACHIEVEMENTS OF WAVE MECHANICS

The Schrödinger Wave Equation embodies three fundamental
assumptions about the nature of matter; that energy is quantised,
that matter has a wave as well as a particle nature, and that the
position and momentum of any particle cannot be known
simultaneously with complete accuracy. For simple physical systems
the equation can be solved exactly; the energies of these systems
prove to have only a limited number of finite values, to each of which
there corresponds a particular probability distribution for the particle
in question.

For systems of chemical interest where a peripheral electron is
associated with a positively charged nucleus, the equation is also
directly soluble. If the nucleus is assumed not to move, the only mass
which appears in the solution is the mass of the electron.

It is comparatively easy to write down the correct Schrödinger
equation for quite complex systems, but no mathematical method is
at present available for solving such equations.

2 THE LIMITATIONS OF WAVE MECHANICS

Systems in which three bodies mutually interact, whether they are
protons or planets, cannot be treated by a mathematically exact
method (H_2^+ is a special case). Thus the distribution of the two
electrons surrounding the nucleus in the helium atom can only be
obtained by the construction of an approximate wave function
(instead of an accurate wave equation) which is then minimised by a
method of successive approximations. The wave function which gives
a minimum value for the energy of the system is taken to be correct.
The more complex the system, the more approximate the wave
function will be, but systems with up to six electrons can be handled
by large computers.

Experience gained with these approximate wave functions leads to
certain semi-empirical rules which can be applied in a non-
quantitative way to systems of chemical interest. These rules on their

own cannot predict the occurrence of novel structures, but they can —once this imaginative step has been taken—be used to assess the likely stability of the new compound. The existence of the $B_{12}H_{12}^{2-}$ ion was predicted to be stable five years before it was actually synthesised, for instance.

3 THE LIMITATIONS OF ANY DEDUCTIVE MATHEMATICAL SYSTEM

Mathematics is an example of a very refined deductive system; a particular set of axioms, whether true or false, can be made to generate a unique branch of mathematics. For example, the geometry of Euclid is derived from the axiom that the shortest distance between two points is a straight line, but there are other self-contained geometries in which this statement is not true. No deductive system of this kind can generate new knowledge on its own, for however elegant the results that are achieved by deductive mathematical manipulations, they contain nothing that was not already implicit in the original axioms. Differentiation is a good example of just such an exact deductive process. It differs sharply from general integration, where the integration constant can only be obtained by reference to something outside the mathematical system altogether—either an imaginative hypothesis or an experimental result.

Mathematics is of enormous importance to science, because it can provide a mathematical model for observed events which can be manipulated by standard procedures to yield deductive results. If an experimental system in the real world appears to behave in an analogous way to a particular mathematical system, it is a good working hypothesis that the deductive consequence of the mathematical system will hold for the experimental system. For instance, once a formula has been found which agrees with the observed rate of rise of a rocket over the first few seconds of its flight, the future progress of the rocket may be forecast by generating values from the formula for longer time intervals.

Wave Mechanics is a very elegant mathematical model whose deductive behaviour happens to correspond exactly with the observed behaviour of a few simple systems. If the computational difficulties could be overcome, it is theoretically possible that like the hydrogen atom, any other GIVEN chemical system could be shown to possess certain finite energy values to each of which there would correspond a particular electron probability distribution. However, this begs the vital question as to what chemical systems are GIVEN—and the only two answers to it both involve human beings: either they were postulated by the imaginative act of a theoretician who fed his hypothetical structure into the Wave Equation, or else the actual structure was discovered in the laboratory by a research worker.

Thus there is no chance that chemistry will ever become a branch of applied Wave Mechanics, however great the computational advances. These calculations may one day take the place of the hack research which classifies analogous compounds after a new breakthrough, but they cannot replace the conceptual insight that made the breakthrough in the first place. All university textbooks on physical chemistry now contain a wave mechanical treatment of the 'sandwich' compounds, but nevertheless they were first discovered as recently as 1951 by a man who was curious about a stable orange sublimate.

Index